W9-CXT-077

ON THE CABLE

ON THE CABLE

The Television of Abundance

REPORT OF THE SLOAN COMMISSION ON CABLE COMMUNICATIONS

McGraw-Hill Book Company

NEW YORK   ST. LOUIS   SAN FRANCISCO

DÜSSELDORF   LONDON   SYDNEY   TORONTO

MEXICO   PANAMA   KUALA LUMPUR   MONTREAL

NEW DELHI   RIO DE JANEIRO   SINGAPORE

42949

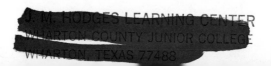

Library of Congress Cataloging in Publication Data
Sloan Commission on Cable Communications.
    On the Cable.
1. Community antenna television. I. Title.
    TK6675.S57   384.55′47   77-37740

ISBN 0-07-058206-8
ISBN 0-07-058205-X (pbk)
2345678910 MUMU 765432

This book was set in Caledonia by University Graphics, Inc.
It was printed on acid-free, long-life paper and bound by
Book Press, Inc. (paperback edition by The Murray Printing
Company). The designer was Christine Aulicino. The editors
were Nancy E. Tressel and Cheryl Allen. Alice Cohen super-
vised the production.

Contents

Preface

Technology and the problems of technological change have long been matters of deep concern to the Alfred P. Sloan Foundation. The publication of this Report on cable communications, however, marks a first venture by the Foundation into the complex arena of technology assessment.

The Foundation's trustees were responding to the clear need for an assessment of the possibilities, both positive and negative, of cable television when they established the Sloan Commission on Cable Communications in June 1970. Their hope was that out of such an assessment could come careful judgments and recommendations concerning the manner in which a powerful new technology should be encouraged to grow.

The Commission's task was a difficult one. Its Report speaks clearly and constructively to the issues that are central to the future of cable communications. The Report also demonstrates that a technological assessment, properly conceived and properly conducted, must deal with far more than the technology itself. Throughout there is an abiding concern with the social and economic consequences of a new technology, and with how the public interest might best be served by its growth and development.

On behalf of the trustees and officers of the Foundation, I extend a special word of thanks to the members of the Commission, to its chairman, Dr. Edward S. Mason, and to its staff. While the Foundation is proud of its role in establishing the

Commission and in assisting it when it could, the Report of the Commission, like all such reports, reflects both the independent nature of the Commission's deliberations and the independent conclusions of its members. It may well create lively discussion and, indeed, some controversy; but if it commands wide interest and thoughtful attention, the expectations of both the Commission and the Foundation will be amply fulfilled.

Nils Y. Wessell   *President,*
Alfred P. Sloan Foundation

Foreword

Since June 1970 the Sloan Commission on Cable Communications has been occupied in exploring some of the problems and possibilities that will arise as wired broadband communications systems spread across the United States. Cable television will expand mainly because it is capable of providing more of the same type of programs now provided by conventional television and of providing better reception in some areas. But if this were all it could promise, the Sloan Commission would not have been established.

The expansion of cable communications can mean much more to the American people than better reception and more conventional programming. This new and immensely powerful technology can also be put to use to serve the public interest in a great variety of ways. It was to examine this proposition and to recommend guidelines to achieve it that we were convened. This Report is the product of our efforts.

We present it with due modesty but also with some confidence. With modesty, because we have come to appreciate the enormous complexity of our assignment. With confidence, because we have been the beneficiaries of a wealth of assistance from a broad array of experts and authorities.

In the appendix to this Report can be found a list of all those who helped inform our deliberations by serving on the staff, preparing special studies, testifying before the Commission, or helping us think through particular problems. We are grateful to them all. I can only mention the most important sources of assistance here.

First and foremost we are indebted to the trustees and officers of the Alfred P. Sloan Foundation, who initiated the undertaking, organized the Commission, financed its work, and provided assistance to the Commission whenever called upon. Our findings, however, are our own.

Secondly, we owe a debt to a dedicated staff led by Paul L. Laskin, director, and Monroe Price, deputy director. They provided the background materials and early drafts of this Report, which helped focus our deliberations and which led us to the timely completion of our task.

And finally, I want to salute my fellow commissioners. They have worked long and hard for nearly eighteen months in an area where many of us initially lacked familiarity or experience. It has been a truly superb group and I am honored to have been its chairman. Together, we have been immensely challenged, occasionally frustrated, never bored. We hope the Report we have fashioned will contribute to the unfolding future of communications in America.

Edward S. Mason

CHAPTER ONE

AN INTRODUCTION TO CABLE TELEVISION

Spreading quietly into every corner of the United States—slowly and unevenly and yet with its own air of inevitability—is a new communications technology. It is known sometimes as Community Antenna Television, sometimes as Cable Television, and as it becomes more familiar simply as "the cable." In being for about twenty years, only recently has it begun to warrant notice on the front pages of the daily press, and what is perhaps more significant, on the financial pages as well. It is moving now from the town to the metropolis. Once a recourse for those who had no other access to television, it is becoming the instrument of choice. Once a supplement to conventional television, it is beginning to be recognized as a threat.

At this moment, it is not remarkably impressive. Cable television is to be found primarily in small cities and towns, unserved or only partially served by conventional television, and in a few large cities where local conditions make conventional television something less than complete or less than satisfactory. Most systems are small, providing far fewer channels of television than the state of the art would permit. All of them together

serve roughly 9 percent of the potential audience. With relatively few exceptions, each system is self-contained: it does not
have the capacity to link with others. And finally, cable television possesses virtually no significant programming in its own
right; it is almost totally dependent upon the products of conventional television.

Indeed, it is not the present reality of cable television that is
of interest. Most systems now in being provide twelve channels
or fewer, but engineers speak confidently of 20-channel systems
that will soon give way to 40-channel systems. Satellite experts
speak of low-cost interconnection that can, upon demand, weld
thousands of independent cable television systems into a single
national system or a dozen regional systems. Economists predict
the day when 40 percent or 60 percent or even 80 percent (depending upon the assumptions they make relating to public
policy and other imponderables) of all homes in the United
States will be "on the cable." Programmers are stirring with
ideas of what they will be able to provide when the day comes
that the system begins to mature.

Cable technology, in concert with other allied technologies,
seems to promise a communications revolution. There have been
such revolutions before. Some 500 years ago the hand-written
manuscript gave way to the printed book, and where earlier
the store of man's knowledge and judgment and imagination had
been available only to a few thousands of the wealthy or the
learned it abruptly was laid bare to all who wished access to it.
Some hundred years ago the first telephone wires were strung,
and where earlier a man could readily make immediate contact
with no more than those persons he chanced to find in his own
neighborhood, quickly he began to find the whole city, the whole
nation and ultimately the whole world within the sound of his
voice. The revolution now in sight may be nothing ·less than
either of those. It may conceivably be even more.

Yet we are more wary today of technological revolutions,
wherever they may take place. We have seen, in the last half

century, a transportation revolution and there are many who are by no means certain that they admire what it has brought. We have seen revolutions in the production of food and the conversion of energy that appear to pose a threat to the very planet upon which man makes his existence. The notion that technological advance is synonymous with progress has lost its old attraction, and there are many who will assert that it is quite the opposite that is true.

It is absurd to make technology the master of man, either by asserting that what technology can do it *must* do, or that what it can do it *should* do. Society has in fact always made choices from among the technological possibilities that lay before it: political choices, social choices, economic choices. But it has not always made those choices with foresight and judgment, and as the technologies themselves become more complex and more arcane, and their consequences more far-reaching and more nearly irreversible, the exercise of foresight and judgment becomes at the same time more difficult and more necessary.

Choice Is Still Possible

Cable television today is at a stage where the general exercise of choice is still possible. If for no better reason than that there is a history of government regulation in the field of television, it remains possible by government action to prohibit it, to permit it, or to promote it almost by fiat. Citizens may still take a hand in shaping cable television's growth and institutions in a fashion that will bend it to society's will and society's best intentions. It is not as yet encumbered by massive vested interests, although that day may be no longer remote. It is not as yet so fixed a part of the national scene, as for example conventional television is, that it appears almost quixotic to attempt to redirect its energies. There is, in short, still time.

And the incentive is more than the intellectual satisfaction of mediating the development of a new technology. A whole family of urgent problems with which this society is confronted can be looked upon, in part or in whole, as problems in communications. Among them might be counted problems in the uses of the political process, in the relationship between government and the governed, in the control of the non-medical uses of drugs and in the delivery of health services.

All those problems, moreover, and others equally pressing combine in a single overriding problem of the inner city. Here are found the special circumstances that arise within a population unfamiliar or uncomfortable with the printed media and consequently largely deprived of any sophisticated means of communication, internally or with the world outside, directed toward the resolution of its own special needs. The problem of the inner city will not be solved by communication alone, but communications may be brought to play a most significant part. If cable technology proves indeed to be the heart of a communications revolution, its impact upon society's most immediate needs might be enormous.

It was for all those reasons that the Alfred P. Sloan Foundation undertook early in 1970 to establish a Commission on Cable Communications, charged with assessing the promises and the problems of the new technology, in the expectation that before the end of 1971 it would arrive at some reasonable set of conclusions and recommendations and make them known to the general public.

What the Commission will be concerned with in this Report lies somewhere between "communications," broadly conceived, and "television," narrowly conceived. It will be defined more usefully as this Report proceeds. Taken as a whole, our Report can be construed as an attempt to define a whole new area within the general field of communications, and to elaborate the steps by means of which it can be approached.

Need for Regulation

Underlying this Report is a set of considerations upon which the Commission found itself in general agreement. Certain aspects of this Report may be more comprehensible if those basic considerations are explicitly set forth.

Television has always been a heavily regulated industry. Historically, the necessity for regulation is a direct consequence of the nature of the radiated television signal. The radiated signal requires space on the electro-magnetic spectrum, and is subject to disabling interference if the spectrum is not properly allocated; there is simply not enough frequency space within the spectrum to permit all to broadcast who wish to do so. Hence there must be allocation through some means. The history has been for the government first to allocate, then intervene in order to assure that the power it transfers with its allocations is not economically, politically or socially misused.

It may appear that with the cable signal, which is not radiated over the air, the case for regulation disappears. For many reasons, however, this is not entirely so. The Commission believes that some regulation continues to be necessary. Regulation can assist in promoting the beneficial uses of cable and encouraging diversity in those uses. It can help insure the orderly growth of cable in a manner that is fair to all those who have an economic interest of one sort or another in its growth. Because in any locality cable television has many of the aspects of a natural monopoly, regulation can be a source of protection for the public — a manner of insuring that subscribers are fairly treated and that there is a continuing national television system throughout the nation. But at the same time, regulation is not only subject to abuse, but brings into being most difficult problems of determining the public interest.

Much of the latter portion of this Report is concerned with the nature of regulation under which cable television should operate. The Commission has approached the subject with the common agreement that regulation is not desirable simply for the sake of regulation. Wherever regulation is prerequisite to the accomplishment of goals we hold to be desirable, we have favored regulation. But unless that clear case has been made we have consistently cast our vote in favor of the operations of the marketplace.

In particular, the Commission has scrutinized the regulatory framework that has grown up around conventional television, and that with time has come to appear to be the natural order of things. We have attempted to distinguish between those regulatory procedures now governing conventional television which are appropriate to cable television, and those that are not appropriate. After enough time has passed, it frequently appears more unnatural to deregulate than to regulate. Nonetheless, wherever it has proved possible we have recommended deregulation, at least as far as cable television is concerned, in many important aspects of cable television's governance.

Thus, the federal government has designed a complex of regulations intended to protect the viability of over-the-air transmission and in particular of its weaker element, the stations in the Ultra High Frequency (UHF) band; we believe that viability is to be sought in the marketplace, subject only to the need to provide a truly national television service, and we believe further that purely technical reasons may argue for abandonment by television of its grip upon the UHF band.

The government has been generally wary with respect to pay television; we regard it as highly desirable. We see the combination of cable television and pay television as a means by which some higher degree of individual choice may be restored to the provision of entertainment and other services.

The government has restricted and set conditions for cable use of over-the-air programming; we recommend removal or minimization of many of those regulatory restrictions and creation of an open marketplace for the procurement of television programs. If this is accomplished, the "distant signal" problem which has so exercised the Federal Communications Commission becomes insignificant.

From time to time, the government has promulgated or proposed rules regarding advertising on cable television; we recommend that no such rules be applied. The cable operator should in general be free to accept advertising or not, as he may interpret his best interests and his relationship with his subscribers.

The federal government has on occasion sought to restrict entry into the business of cable television; we accept the necessity for regulation, but recommend that at least some of the restrictions be removed. We recommend also that under certain circumstances preference in the granting of franchises be given non-profit and profit-making institutions representing community groups.

We have recommended also an important regulatory role for the states—a role which the states do not now play. We will argue in a later chapter that considerable state supervision of franchising authorities is necessary to a healthy cable television system. We do not believe, however, that state regulatory authority should be assigned to Public Utility Commissions or similar existing bodies.

Common carrier operation of cable television, in which ownership of cable systems would be rigorously separated from control of content that passes along the systems, is often put forward as the most efficient manner in which to assure an extensive nationwide cable television structure. Without precluding the possibility that common carrier operation may some day be advisable, we argue that at the present stage of cable television growth, and for the immediate future, the rigorous separation of ownership and control over content would hinder rather than assist the development of cable television.

Wherever communications are under consideration, there arise questions of libel, obscenity, incitement to riot, and sedition. Because television is, by its nature, so powerful an instrument of communications, there is constantly the temptation to formulate special legislation expressly for the purposes of controlling abuses of the medium. The Sloan Commission has in general taken the view that additional special legislation with respect to cable television is unnecessary, and might in itself constitute a threat to civil liberties; we believe that reliance on existing laws and statutes, the product of centuries of interaction between government and the media of communication, is by far the preferable course.

Hazards of Prediction

The Commission, by the very nature of its charge, has been asked to predict, and to prescribe for the future. We have chosen to interpret that charge to signify the relatively near future, and we have attempted to make both our predictions and prescriptions open-ended.

The Commission simply did not believe that its own collective wisdom—or indeed that any collective wisdom—was sufficient to justify apodictic statements about the nature of broadband communications or of communications in general in the year 2000 or at some other remote moment in time. By the time our deliberations were well along we had the sense that we were beginning to see with some degree of clarity what might happen, and what might reasonably be constrained to happen, during the first half of the decade of the seventies. With somewhat less assurance, but without apology, we believed also that we could make recommendations that might well be worth heeding during those years and the five or ten years that will follow; our range of vision, we were bold enough to believe, was reasonably sound over ten years and perhaps fifteen. Beyond that we were reluctant to venture.

The Commission has attempted to make its recommendations such that they can be readily modified as circumstances alter, and so that actions taken as a consequence of our recommendations will not at some future time become a strait jacket to progress and change. If there is one thing certain, it is that the technology of communications will continue to evolve, and that thé demands society wishes to make upon communications will evolve with the technology. Over the years since the end of World War II, television regulation has been such as to impose upon the United States a system that is far less than the technology of television could now provide. Something of that is inevitable, but the Sloan Commission has at least attempted to make recommendations consistent with the present capability and potential growth of the technology.

The consequence of that decision is a Report that hews closely to the state of the art as it now exists or as it can confidently be expected to develop. It is tempting to venture into the blue sky of technological imagination, and to write knowledgeably about widespread low-cost two-way point-to-point television systems; about home communications centers which can deliver printed copies of any volume in any library on any continent; about coaxial cable systems which will cook dinner, wash the windows and tend the babies. It is salutary to realize that most of those accomplishments may be possible one day, and that day may be sooner than one thinks. But the hard facts of technology, wedded to the even harder facts of economics, provide no warrant for the belief that any of them will come to fruition upon a time-scale that can confidently be established in advance.

Some speculation, however, has been inescapable, and indeed lies at the very heart of this Report. There *is* a new communications system developing, and if the recommendations of the Commission are followed, or if recommendations like them in spirit are adopted as guides to action, it is likely to develop quickly. In making our judgments upon the form such a system should take, we have been obliged to speculate upon the uses to which such a system might and will be put.

Speculation—informed and reasoned speculation, but specu-

lation nonetheless—is almost the only tool the Commission
had at hand for estimating the size and scope of these new uses.
In the end, the shape they will assume will be determined on
the one hand by entrepreneurs, public and private, who are
willing to take the responsibility for risking money and career
on the promotion of an idea or an ideal, and on the other hand
by the users of the system through the response they make to
the undertaking of the entrepreneur.

The strength of such a mode of development is that it does not
depend upon the judgment of one man or of one small group of
men. The outcome is determined by the collective judgment of
the whole community, worked out in the marketplace by measurable success and measurable failure. Tens of thousands of
men and women are thus enabled to contribute to the final
resolution of the problem, and the major contributions are quite
likely to be made by those whom it would be impossible to
identify in advance whatever the process of selection.

The Commission has been forced to guess at outcomes, in the
certain knowledge that it will be wrong in detail, and that it
will also be wrong, at least at times, in its general conclusions.
The subject is in all its elements too complex to justify hopes
of any greater success. We have attempted to look into the future
on the basis of what can be surely determined in the present,
and of what we believe we know about the way rational men
behave, and of what we believe society desires to and should
obtain from a communications system.

CHAPTER TWO

THE TECHNOLOGY— A PRIMER

A television camera is a device that accepts information concerning the scene to which it is exposed, including the sound that accompanies that scene, and encodes it in the form of an electric current. A television receiver is a matching device that decodes the electric current and converts it into a succession of pictures on a cathode ray tube, and once again the accompanying sound. It takes a good deal of electronic circuitry to accomplish all this, but machinery to bring it about, at either end, can be purchased for a few hundred dollars, or less. What remains is the necessity to transmit the signal from the camera which creates it to the receiver which decodes it.

One method employed conventionally is radiation of an electromagnetic wave from a transmitting antenna. The wave travels outward from the antenna—it is "broadcast" in the root sense of the word—losing energy as it travels. But for a distance of about fifty miles or so it can be intercepted by any receiver that lies in its path and is turned to its wavelength, and for another fifty miles or so it can be intercepted (if the radiating and receiving antennas are high enough off the ground) by the

11

use of increasingly elaborate equipment. By that time, the curvature of the earth begins to block the receiver from the transmitter, and the signal is gradually diminished and lost.

Another method, equally conventional, is the transmission of a signal (or many signals) along a wire or a conductor of some other sort. This second method is quite as old as the first. Transmission by wire is, in fact, inseparable from transmission by radiation, for the signal must travel at the outset by wire or cable from the camera to the transmitting antenna, and once arrived at its destination must travel once again by wire from the receiving antenna to the television set. The system in common use is in principle a mixed system, which uses a little cable at the transmitting end, a little wire at the receiving end, and a great deal of electromagnetic radiation in 'between. In practice it can be considered a radiated system.

What Is a Coaxial Cable?

The pure cable system, to the extent that it is a self-contained entity, dispenses entirely with electromagnetic radiation. At its heart lies the coaxial cable which is in part described by the word "coaxial" itself. As Figure 2-1 indicates, the coaxial cable consists of a small diameter inner conductor, a larger diameter outer conductor, a plastic foam to keep them apart and to maintain an electric field between them, and an outer sheath to protect the entire cable from the weather or whatever else might affect the operation of the system.

Such cable can be used to transmit electrical signals from zero frequency (direct current) to frequencies of several billion cycles per second. The coaxial cable used to transmit television signals carries all frequencies between 40 million and 300 million cycles per second. (A telephone wire, by contrast, transmits frequencies between 300 and 5,000 cycles per second.) Since a television signal requires a bandwidth of 6 million cycles per second, a coaxial cable can carry, in principle, the equivalent

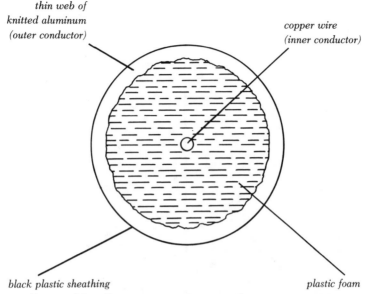

FIGURE 2-1.   CROSS SECTION OF A TYPICAL COAXIAL CABLE

of forty channels of television (although other considerations currently reduce this capacity in practice by about half, to twenty channels of television).

Because the cable must perform the entire task of carrying the signal from its point of origin to the set which is to receive it, there must be a physical cable link between the source of the signal, called the "head-end" of the system, and each subscriber on the system. Figure 2-2 shows how this is ordinarily effected, in a moderately densely populated area, and in its simplest form. From the head-end, a trunk-line runs out through the area to be covered. Feeder lines fan out from the trunk, in a fashion that brings at least one feeder line within approximately 75 to 150 feet of each residence in the area. From the feeder lines, drop lines can be connected directly to each set within a residence. Thus each home is physically linked within the head-end: it possesses a cable connection of its own. Any

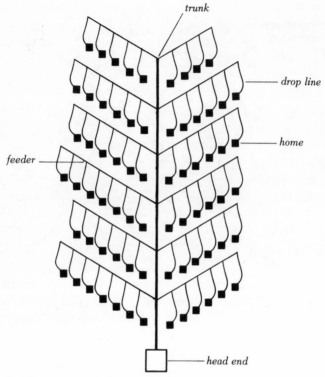

FIGURE 2-2.   SCHEMATIC DIAGRAM OF A TYPICAL CABLE TELEVISION
SYSTEM

signal emanating from the head-end may be received at any
home.

The task is not complete, however, when the cable is in
place. An electrical signal necessarily loses strength as it passes
along a conductor, and amplifiers must be inserted along the
line to compensate for the loss; other amplifiers are necessary
as bridges between trunk lines, feeder lines and drop lines.
Initially vacuum tubes were utilized, but newer amplifiers employ solid state technology and indeed the whole field of
amplifier design and development is in flux. Rapid growth

of cable television will inevitably mean rapid development of the associated electronic technology.

It is the capacity and the cost of amplifiers that limit at present the capacity of the cable. Although existing coaxial cable in theory will carry forty channels, that theoretical limit cannot be met with existing amplifiers. It is, however, possible to lay cable in a fashion that enables the system operator to replace amplifiers as new designs become available; thus the same cable that today carries twenty channels or less may at some future date be adapted to carry the full load of forty.

Other electronic circuitry is necessary along the cable if it is to be fully utilized. Filters which block out part of the band make it possible to deny a household access to one or more channels; such filters might be necessary if, for example, a special channel were to be reserved for the medical profession. Electronic valves governing the direction in which the signal flows would make it possible to arrange for "up-stream" use of one or more channels; that is, transmission of a signal from a designated point or points back to the head-end.

Up to this point in our description, the system is without the television signal itself. This is fed into it from the head-end, which can provide programs in three fashions. The simplest requires nothing more than the erection of an antenna, which will pick signals out of the air from conventional television stations operating in the vicinity, and transmit them along the cable: the more elaborate the receiving antenna, the larger the "vicinity" and the more stations made available. If, however, a desired station is beyond the range of even the most elaborate antenna—if, for example, a cable operator in Las Vegas, Nevada, wishes to pick up a signal from Los Angeles—he need only erect his antenna somewhat closer to Los Angeles than Las Vegas and transmit the signal the rest of the way by means of a long-distance micro-wave or cable link, which can be rented from American Telephone and Telegraph or (under recently promulgated FCC rulings) from a competing operator. Finally, the cable system operator can set up, at his head-end television studio, facilities as elaborate as he chooses to pay for, ranging from a simple video-tape machine to a full studio with its complements of color television cameras, lights, monitors, and all the accompanying paraphernalia.

Limitations of Radiated Television

The transmission system that has just been described—cable television—has from the beginning of television existed side by side with the radiated system. Yet it was all but inevitable that television should be born and come to maturity as a radiated system, rather than a cable system. It was natural, in the first place, to construct it after the model of radio, for as a technology it was merely an extension of radio. But there were even more pressing reasons. Comparing the cable-carried signal with the radiated signal in any given area, one persuasive criterion of choice between them was the difference between $5,000 a mile over thousands of linear miles—the minimum cost of laying cable—and virtually no cost at all.

In the early days of television, when receiving sets were few and revenues from the sale of time relatively small, the expense of laying cable could not conceivably have been borne by the television entrepreneur or by any resources he was able to command. In contrast, a transmitting antenna was relatively cheap to build and operate; it had the further advantage that no subsequent investment of capital was required as new receiving sets came on the line.

Yet there was a price to be paid, for there were from the outset deficiencies in radiated television, and they had a profound effect upon its development. Indeed, it can be argued that those deficiencies created not only the physical configuration of the television system the nation now enjoys, but the nature of much of the programming that passes over that system.

The shortcomings arise out of aspects of the physical laws that govern the radiation of electromagnetic waves. First, only waves with certain well-defined limits of frequency—from approximately 50 million cycles per second to 200 million cycles per second (see Figure 2-3)—are eminently suited for television transmission. That is known as the Very High Frequency (VHF) band. Within it must be found room for FM radio and for other services; what is left provides space for only twelve television channels between 54 and 88 million cycles per second, and between 174 and 216 million cycles per second. Above the VHF band lies the Ultra High Frequency (UHF) band with room for an additional seventy channels, but the natural laws of electromagnetic propagation make these less desirable for television as one proceeds up the range.

A second physical law is that waves at the same or very nearly the same frequency interfere with each other as they radiate through space, and interfere even if their sources are separated by greater distances than one would expect from the limited range of useful signals. As a consequence, the presence of an antenna radiating a signal on Channel 3, for example, makes it impossible for another station to transmit at that frequency within a radius of approximately 200 miles, and additionally makes it impossible for another station to transmit on either Channel 2 or Channel 4 within a radius of approximately 100 miles. Interference from neighboring channels and man-made noise is even more marked in the UHF band.

In the early days, scarcity of space in the electromagnetic spectrum and the necessity to separate stations operating at the same or neighboring frequencies made it necessary for the FCC to enforce a freeze on the construction of new stations while it came to terms with the allocation problem. Clearly, since there was a limit to the number of transmitters that could exist in the continental United States without destructive mutual interference, it was necessary to allocate those stations geographically in the best public interest. The FCC wrestled with

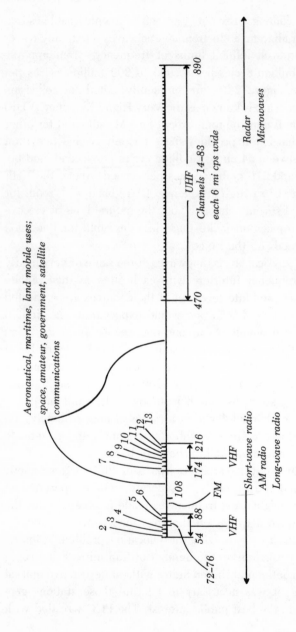

FIGURE 2-3. THE TELEVISION SPECTRUM

18

the problem for four years, and in 1952 emerged with the pattern of allocation under which the television system now operates. It is not a simple pattern, since the problem to which the FCC addressed itself had no simple solution. But in effect what the FCC attempted to do was provide as many people as possible with minimum service — at least one channel — and provide in areas of progressively greater population-density progressively greater service. Today, 97 percent of America's households receive at least one channel, more likely than not a VHF channel. At the other extreme, Metropolitan New York City, with the greatest population density in the country, is served by seven channels in the VHF band alone, the maximum that the VHF band permits. Los Angeles, a densely populated area largely surrounded by unpopulated areas, is also able to enjoy seven VHF channels. Chicago has five VHF channels; St. Louis, five; and Boston four.

The following tabulation shows the outcome of the FCC allocations:

NUMBER OF CHANNELS	PERCENT OF POPULATION SERVED
10 or more	17%
9	9%
8	11%
7	20%
6	9%
5	13%
4	11%
1–3	7%
0	3%

There are, however, concealed inequities in the table as given. Slightly more than one-third of all television stations now in operation transmit in the UHF band. Inherently, the UHF channel is inferior, for the higher one moves in the electromagnetic spectrum, the more the waves tend to travel in straight lines, reflecting off obstacles rather than bending around them, and the more energy the waves dissipate as they travel out, restricting their range. There have been technological problems, moreover, associated with both the dissemination and the reception of UHF signals, and although they have been largely overcome they plagued UHF television in its early days. The result has been that commercial television has grown up largely as a VHF system. The networks are predominantly on VHF; sets (despite FCC attempts to intervene) are primarily VHF sets. Although the FCC has issued a total of 584 construction permits for commercial UHF stations, 315 were later canceled; 54 percent of the permits, that is, were never taken up or the stations built. Of the other 269 permits, 186 commercial UHF stations are currently in operation and of those which exist as independent stations (without network affiliation) only a handful operate at a profit.

Demands on Spectrum Space

Television thus came into being constrained by the general scarcity of space within the electromagnetic spectrum, and constrained further by the even greater scarcity of VHF space. But the problem of the electromagnetic spectrum has further dimensions.

The portion of the spectrum available for television lies broadly between 30 million cycles per second and 1,000 million cycles per second, with the lower portion of the spectrum the more desirable. Below 30 million cycles (which in any case is not much spectrum space) propagation characteristics are such that a satisfactory television picture is unlikely. Above 1,000 million cycles, waves begin to behave more and more like light waves, and require a clear and unimpeded line of sight between transmitter and receiver.

The characteristics that make 30 to 1,000 million the band of choice for television make it also the band of choice for other important uses. In particular, wherever communication is neces-

sary between sources of which one or both are in motion during
the communication, or between sources of which one or both
have no relatively permanent fixed location, that portion of
the spectrum is the instrument of choice. The communication
itself may be relatively trivial, as in the case of most Citizens'
Band uses; it may be of pressing importance, as in the case of
communications between aircraft and ground control installations, or maritime vessels and shore installations; it may be
somewhere in between, as in the case of taxicabs and their dis-
patchers. It may also be somewhat esoteric, as between radio-
astronomers and distant stars, or space vehicles and Houston,
or satellites and ground stations. Whatever it is, it makes demands on the spectrum space that might otherwise be used,
more or less efficiently, by television.

Those alternative demands are great, and are growing; moreover, they will continue to grow. To all those who have need
for spectrum space—and the United States government itself,
with its security obligations, is foremost among them—the 53
percent of the most useful part of the spectrum that is now
occupied by television in one form or another is a target. If
other uses are to be served, the dominance of television within
that part of the spectrum must be whittled away.

And they can point to one essential difference between television use of the spectrum and its use for mobile services: television now has a choice between radiated television and cable
television; the mobile users have no such choice. It has become,
with technological advance, relatively easy to lay a cable between transmitter and receiver. There is no likelihood that one
will be able to lay a cable between a ground station and a Boeing
747, or army headquarters and a battalion in the field, or Houston and the moon. Over most of television's activities, its pre-
emption of scarce spectrum space is unnecessary, and in the
face of other pressing needs is accordingly indefensible; this,
at least, is the extreme position that a deprived applicant for
spectrum space might maintain.

That is the technological case for cable television: from the point of view of technology alone, a large part of television should be transferred to cable, leaving radiated television only for those instances where cable will not serve and liberating virtually the entire electromagnetic spectrum for those services to which it is indispensable and which are clamoring for it. Some of the spectrum would be retained for television, so that remote areas and portable television sets might still be served. But the lion's share of the spectrum would go where it is needed, and where no alternative exists.

We are inclined to believe that the growth of cable television will lead sooner or later to the abandonment of at least the share it now commands of the UHF band in the electromagnetic spectrum. We believe further that the best use of the spectrum, in the light of all the demands made upon it, together with the capacity of cable television to fill needs for television service, makes such an outcome desirable, and we would favor rather than oppose steps to hasten the change in spectrum allocation among competing uses. For the foreseeable future, however, we see no imminent threat to television occupance of its present share of the VHF band.

In any event the case that can be made by technology has served primarily to reinforce the Commission's confidence in its own case for cable television, which rests on other grounds. The theme developed in this Report is that television, so long as it remains bound to electromagnetic radiation, is underused, and that the American public is thereby underserved. The limits of the spectrum, even when most of it is given over to television, impose limits on television that are no longer necessary. It was precisely this fact that gave birth to cable television.

CHAPTER THREE

A BRIEF HISTORY

In every sense except the geographical, Palm Springs is a suburb of Los Angeles; it looks to Los Angeles for most of its services, and its life-style is patterned on that of Los Angeles. But as far as television is concerned, it is cut off from Los Angeles by a mountain range. There arose, accordingly, a demand in Palm Springs for access to Los Angeles television, and as usual, there were those who were anxious to fill the demand. An entrepreneur erected, on top of a nearby mountain, an elaborate receiving antenna. He then offered, for a fee, to connect any householder in Palm Springs by means of coaxial cable to his antenna. At once, for a few dollars a month the seven VHF channels and three UHF channels from Los Angeles were at the disposal of anyone in Palm Springs willing to pay the price.

In other areas, the incentive may initially have been different but the consequences were the same. In rural areas the local radio dealer sensed a demand for television that was being stifled in the absence of a television signal. He, too, invested in an elaborate antenna, set on a high point of ground, and strung his cable. His returns were partly in the monthly fee he charged, and partly in sales of television sets.

23

These were the Community Antenna Television services—
CATV—with which cable television began. They were almost
without exception small enterprises, conducted locally and
providing a purely local service. They provided television where
otherwise there would have been no television. They added to
the audiences of the television stations they imported, without
damaging in any way those stations or any others. There were
many such CATV[1] services—by 1960 they were estimated to
number 640—but they were of purely local consequence, and
left television as a whole unruffled.

While all this was going on, television was growing in a
fashion that no one, as recently as ten years earlier, could
possibly have predicted. Television had become a predominant
feature of the American scene, the primary instrument to which
the mass of the population turned for entertainment and information.

Cable Invades San Diego

The voracity of the appetite for television proved to be im-
measurably greater than anything that FCC allocations could
satisfy. And thus the next stage in cable television was reached.
Again, the example can be taken from California: this time San
Diego.

By any of the criteria that the FCC had adopted in 1952,
San Diego was well served. Within the city limits were two
VHF stations, one of them affiliated with NBC and the other
with CBS, both with antennas on high ground and providing
service to the entire area. A few miles away, outside FCC
jurisdiction in Mexico, was a third VHF station affiliated with
ABC. The three together provided what could certainly be construed as a full diet.

---

[1] CATV has become an obsolescent term in that "community antennas" are
in decreasing use, least of all in new systems, to bring signals into the head-end.
The term, CTV, standing simply for cable television, is much to be preferred.

Yet, in 1961 cable television invaded San Diego. The entrepreneurs erected a sophisticated antenna, capable of picking up Los Angeles channels from 100-odd miles away, and for a fee of $5.50 a month after a modest installation charge of $19.95 — not always insisted upon and later substantially reduced — offered full Los Angeles service to San Diego viewers. Since San Diego was already receiving the three networks, what was being offered in fact was the four independent stations that served Los Angeles with sports, old moving pictures and reruns of network shows, plus the local Los Angeles services provided by the three network affiliates. That was enough. By the end of the decade, the San Diego system was the largest cable television system in the United States, serving 25,000 subscribers. And all this in a city that provides perhaps more opportunities for non-television recreation than any city in the United States of comparable size.

San Diego demonstrated that three channels are not enough to satisfy an ordinary audience, and that a large part of that audience is willing to pay cash out of pocket for more diversified programming than the three networks by themselves provided. At the same time, other communities with modest channel allocations were demonstrating that the system of cable television was not necessarily limited to what a costly antenna could pick off the air. In various locations, where even the extended radius of reception provided by expensive community antennas was insufficient, enterprising business men leased micro-wave links from American Telephone and Telegraph to lengthen their reach, bringing stations into the community from more than a hundred miles away and retransmitting them to their subscribers by cable. The uneven pattern of the FCC allocation, in short, was being filled in by independently controlled community antennas and where these were insufficient by specially leased micro-wave links. For the country as a whole, the system had become a mixed radiated and cable system, even though cable was by far the junior partner.

Add Color and Program Origination

Meanwhile, the spread of color television was introducing a new element. Even VHF television signals, unlike radio signals, tend somewhat to bounce off large obstacles rather than bend around them. Thus a tall building reflects the signal, and in doing so acts like a weak transmitter, rebroadcasting the signal at the same frequency as the station from which the signal originates. The result is interference, just as if there were two stations broadcasting in the same region at the same frequency. Because the reflected signal is weak, the interference is not pronounced. On a black-and-white set it is usually barely noticeable, although at worst it produces ghost images. But with color reception, far more sensitive to interference, the result is quite likely to be a quite unsatisfactory picture. In New York City, for example, with its multiplicity of tall buildings and the consequent multiplication of reflected signals, color reception can be very bad indeed.

Here was a new opportunity for the cable television entrepreneur. He was able to offer, in New York, something that radiated television could not always supply: a high quality color picture. Cable television invaded New York, despite the presence of a full complement of VHF channels, and began to attract subscribers. New York City franchises became a prize for which the major cable television companies were anxious to compete.

And quickly, the enterprise took on a new aspect. In New York, unlike elsewhere, what the cable system had to offer was not a new picture but merely a better picture. The inducement was smaller, and in marginal situations clearly not enough. Rather than spend $5.00 a month, most New Yorkers within the franchise area were quite willing to put up with a picture that was slightly less than perfect. Something more was needed.

The something more turned out to be programming that was of high interest, and that was not available either on network television or on independent television. Specifically, the systems within New York City sought out exclusive rights to home games of the local basketball and hockey teams. (As in conventional television before it, the first impact of the new system was made by way of sporting events.) In addition, special programming was provided for the black and Spanish-American enclaves that existed within the franchise area.

Cable origination, as it is called, was not entirely new. Most cable systems, even the smallest CATV systems, conventionally used an otherwise unused channel or two by permitting an open, untended camera to transmit news directly off a ticker, or weather off the faces on an instrumental panel. A few systems transmitted low-cost local programming, usually prepared and performed by amateurs or high school groups; a few had even gone so far as to transmit local amateur athletic events. But what New York provided, for the first time, was programming at the level of over-the-air programming, available to cable subscribers alone. The dependence of the cable upon over-the-air transmission was no longer complete. For a few dollars a month one could buy what the rest of television could not provide: athletic events or neighborhood programming of major interest. In New York today, the growth of cable television is limited for the moment primarily by the ability of the cable operators to lay cable and to merchandise their service. The two systems in operation were serving, by mid 1971, more than 80,000 subscribers.

And with that, the prehistory of cable television (one might call it) has come to an end. It had begun as a substitute for over-the-air television where over-the-air television did not exist. It grew later as a supplement to over-the-air television. Today, it has proved that it can be a complement to over-the-air television, providing services that networks and independent stations do not provide. There is only one stage remaining to it: as a replacement for over-the-air television. It is not impossible that it will some day reach that stage.

Cable and the FCC

As the penetration of cable increased beyond the filling of otherwise empty television space, the Federal Communications Commission found itself obliged to grapple with the problems it was beginning to create. Initially the FCC had taken the view that CATV required no special regulation: it was merely doing what the television system would otherwise be unable to accomplish, and doing it without any noticeable interference with established interests. But as cable television began to make its way into communities that were already served, however minimally by established stations, the FCC became more sensitive to the issues that were being raised.

In particular, the FCC has been sensitive to the threat of cable television to the UHF stations it has licensed, and to which it has been committed as a solution, however imperfect, to the problem of channel scarcity. Over the years that commitment has become extensive, culminating, in 1962, in rules requiring that all television sets be capable of receiving UHF as well as VHF signals. (In the absence of real demand many manufacturers have combined quality VHF capacity with inferior UHF capacity, and the intent of the rules has been in part evaded.) The UHF independent station, struggling for survival, may gain initially when cable penetrates its territory and begins to retransmit the UHF signal in a manner that brings it in more crisply and more readily, but in the end cable becomes one more competitor in a market where UHF is already at a disadvantage.

But even the local VHF station is affected. In a small city with but one VHF station, that station gets all the viewers and collects advertising revenue accordingly. Bring in two network competitors by cable and at once its share of audience diminishes by something approaching two-thirds — and its revenues diminish accordingly.

The FCC felt constrained to act, even though it was not entirely clear that it had the authority to act, and struck at cable television at its weakest point: program access. Almost without exception, cable operators held out as inducement to new subscribers what have come to be called "distant signals"—programs broadcast by stations over the television horizon, and sometimes hundreds of miles away from the cable installations. In 1966, the FCC imposed a ban upon the importation of distant signals into the hundred largest television markets, in which are to be found nearly 87 percent of the American viewing public. The pronouncement of that rule led to a judicial challenge of the FCC's authority to regulate cable; that authority was upheld by the Supreme Court in 1968, in *United States v. Southwestern Cable Company.* The result was a "freeze" in the cable industry, at least in those areas where most Americans live.

In addition, copyright owners of the imported programs, and in particular the motion picture industry, were becoming concerned as cable operators profited from copyrighted material without payment of fee. Existing copyright legislation dates back to 1909, when neither conventional nor cable television needed to be taken into account. Litigation proceeds in circumstances where the outcome is determined by judicial interpretation of outdated legislation; the FCC has urged Congress to provide new legislative guidance, but the entire issue has been caught up in Congressional cross-fire and remains unresolved.

As the 1960's came to an end, the FCC began to reflect a new attitude. The broadcast industry was not quite so monolithic in opposition to cable, for some broadcast interests were beginning to venture into cable. As cable spread, its public constituency grew, and frustration with stringent rules on importation of signals was freely made known. The lure of cable grew; people began to talk, and journalists write, of the potential value of cable to towns and cities, of a communications revolution, of a medium which might have an important impact on informational and cultural processes.

By 1968, the FCC was trying to free itself somewhat from the restrictive posture of the past. Most recently, in August 1971, it transmitted in a letter to Congress broad outlines of rules under which the FCC proposes to proceed to govern cable television; among other provisions, those proposed rules open up the top hundred markets for the importation of distant signals. The rules are subject to full discussion by all those concerned, who may appear before the FCC or file comments. The FCC hopes to conclude the process by issuing its definitive rulings by March 1, 1972, a date which it will be hard put to meet and which can be delayed by Congressional intervention. Whatever the rules that may be adopted, their impact will be governed by the resolution of copyright problems, either by judicial decision or by new legislation.

But in any case, there appears to be a general concession that the system is here to stay, and even a general intention to encourage its growth, at least mildly. It is reasonable to believe that cable television has at the very least an amber light ahead of it, and in all likelihood a green light. It is substantially free, or promises to become so, to develop in some degree in accord with its own imperatives.

Shape of the Cable Industry Today

As the system stands today, in the fall of 1971, it is still small and still financially unrewarding, at least as compared to conventional television. But it is by no means insignificant, nor is it entirely without financial resources.

In the ten years preceding January 1, 1971, as illustrated in Figure 3-1, the cable television industry grew from 640 systems serving 650,000 subscribers to 2,500 systems serving 4.9 million subscribers; its annual growth in subscribers took place at a compounded rate of 22 percent. These figures include a period, in the late 1960's, when growth was severely inhibited by the entry of the FCC into regulatory activities, most of them inimical

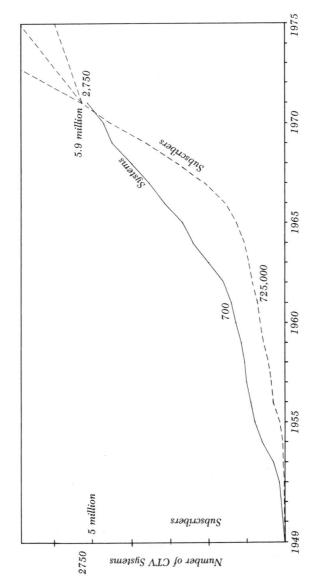

FIGURE 3-1. HISTORICAL GROWTH OF CABLE TELEVISION SYSTEMS

to cable television. Even during that period, however, subscribers continued to increase at the rate of 22 percent. At that
rate, it might be noted, if it should continue over the decade of
the seventies, penetration of the system by 1980 will be in excess
of 50 percent.

The months since the beginning of 1971 have seen continued
rapid growth, particularly in New York City. As closely as a
count can be made, there are now in being 2,750 distinct cable
systems. They range in size from small CATV systems serving
isolated rural areas to the giants in New York and San Diego.
In all, the systems now reach 5.9 million households, or approximately 9 percent of all television households in the United
States. What is perhaps more significant, franchise applications
have been granted, or are being entertained, in more than half
of the top thirty markets in the United States, among them New
York, Philadelphia, San Francisco, St. Louis and Atlanta. Construction is actually in progress in cities, small and large,
throughout the country; the National Cable Television Association estimates that new subscribers are being added at the rate
of about 80,000 a month.

Many systems are still what the industry itself calls "ma and
pa operations," built by local individuals or companies with
modest assets and modest goals. Indeed, over the country as a
whole the average system has approximately 2,000 subscribers.
But in recent years, larger and more substantially financed companies have entered the field; one such company, Teleprompter
Corporation, has now more than 600,000 subscribers, or about
one-tenth the entire industry. Others are approaching and even
threatening to pass that number.

Approximately half the cable systems in the United States
are owned or controlled by companies that are otherwise engaged in communications. Over-the-air broadcasters own 30
percent, newspapers and publishers 12 percent, telephone
companies 5 percent. One network, the Columbia Broadcasting
Company, had extensive interests in cable television before it
was obliged by an FCC ruling to divest itself of them.

The cable industry as a whole has not been distinguished by a high level of effort in research and development. The operators themselves do almost none. The industry, moreover, has over its lifetime not been large enough nor well enough financed to encourage research and development on the part of equipment manufacturers; the market was not large enough to support anything more than limited improvements and marginal innovations. To some extent, this deficiency will correct itself, at least as far as immediately perceived needs are concerned, with the growth of cable television.

Over the system as a whole, there is little important origination, although the New York experience is setting a pattern that will certainly be imitated as the remaining top markets are wired. Except in New York subscriptions are still being sold primarily on the basis of additional over-the-air services and a crisper color picture.

There is little interconnection among local systems, and that of minor importance. But there are important stirrings in the field of interconnection. Both COMSAT and Hughes Aircraft Corporation have plans to interconnect by satellite; in the case of Hughes the proposal incorporates plans for program origination.

Commercial interests are becoming progressively more interested in the potential of cable television for the provision of services other than entertainment and information. The same cable that brings ordinary television fare into the home can be used as a conduit to conduct ordinary business: to buy and sell, and to operate in the transfer of payment for merchandise. With some elaboration, it can be arranged to move data of all sorts in and out of the home. The cable, properly arranged, can read meters, serve as a fire and burglar alarm, make market surveys, and even conduct political polls. None of this is fanciful: such services already exist experimentally or are being actively planned.

Finally, as rudimentary as it still is, the system appeals to the American love of gadgets. For a great many people, a few dollars a month is by no means too much to pay for a black box with which the proprietor may divert himself, bringing to his television set something that may be little more in substance, but still more than his neighbor can receive.

In short, there is every sign that cable television, even as it is now constituted, will continue to grow. What the growth can mean to the American people, and how it can be assured, is the substance of this Report.

CHAPTER FOUR

PROSPECTS
AND POSSIBILITIES

The cable system that now exists, rapidly as it may be expanding, is readily understood. To abandon the firm ground of accumulated data and venture across the line that separates the present and the future is a somewhat more perilous task.

The Commission has sought guidance from engineers, economists, and communications experts, to help it come to some understanding of the likely development of the cable television industry. We have attempted to determine, at least to our own satisfaction, how rapid the future growth of cable television is likely to be, how deeply cable television will have penetrated in the years directly ahead, and what will be the likely technological conformation of the cable system as it grows toward maturity. The full impact that cable television will have on American society, if it is to be different in any substantial way from the impact of conventional television, depends to a large degree upon the answers to those questions. But the questions themselves are not entirely distinct. There is reason to believe, as we shall point out below, that at some point the rate of growth of the system will leap ahead, until it reaches an upper limit.

There is reason also to believe that the development of technology, and in particular the technologies of cable capacity and intercommunication, will also profoundly affect the rate of growth.

There is one important contingency. As we have reiterated, wherever the system is growing, it feeds on its ability to provide within its area something more than conventional television can provide. That something may be more signals, different signals, or better signals, but fundamentally the incentive to make monthly payments is provided by a deficiency, real or imagined, in over-the-air service. It is not impossible that the capacity to provide greater diversity may be regulated out of existence. In the present state of cable television, copyright legislation expressly designed for the purpose might conceivably strangle cable television in its infancy. Regulations prohibiting local origination — and such regulations have been suggested — might effectively keep cable television out of the major metropolitan areas, and condemn it to a permanently secondary status.

Our Main Predictions

The Commission has reached conclusions concerning the likely development of cable television during the decade of the seventies. Those conclusions rest on two basic assumptions. The first of the two is a plausible assumption: that the technology of cable television is not likely to alter radically during the period, and the bold outlines of the system will be much as they are today.

The second assumption, is that the recommendations of this Commission will be heeded and that no seriously restrictive legislation will be passed by Congress, and no seriously restrictive regulations issued by the FCC. Despite the current relaxation of FCC views, this is by no means a foregone conclusion.

Given the potential impact of cable television, and the consequent opposition it is certain to arouse—indeed, has already aroused—among competing media, the assumption is by no means entirely safe. But the Commission is also convinced that cable television is here to stay and is almost certainly bound to grow, and that even the imposition of strong restrictions is more likely to affect the time-scale in which cable television develops than the end result.

In the context of those assumptions, the Commission anticipates the following:

1. *Channel capacity.* The number of television channels that a coaxial cable can carry is a function of the size of the cable and the sophistication of the associated electronic equipment. The earliest systems were capable of carrying six channels into the subscriber's home; later systems raised the number to twelve, conforming to the unadapted capacity of the VHF television receiver. More recently, the capacity of new installations is of the order of 20 to 25 channels, although even now a few forty-channel systems are under construction or contemplated.

The Commission is advised by its consultants that by the end of the decade the majority of cable franchises will have a capacity of at least twenty channels, that forty-channel systems will be commonplace, or at least well within the state of the art, and that even greater capacity may be found in great metropolitan areas. The Commission is further advised that this is a conservative prediction; that it is at least conceivable that ordinary channel capacity will rise to eighty or above by the use of paired cables or more capacious cables. So far as reasonable prediction is concerned, even twenty channels will provide, for the average householder at least, as many television signals as he can possibly use, and in all likelihood a great deal more than he can use. As will be seen when this Report turns to the question of uses, this massive capacity, and indeed the existence of excess capacity, is of great significance.

2. *Penetration.* The willingness of the householder to purchase a subscription to cable television depends directly upon the extent and nature of the new services it will provide to his household. In rural areas, the provision of one, two or three new channels may be inducement enough; in large metropolitan areas it may be necessary to provide wholly new programming, unobtainable from conventional television. But since the economic base for cable television, like conventional television, is fixed by the absolute size of its audience, there is a converse proposition that is also true: the ability of cable television to provide new services, particularly in the large cities, will depend directly upon the penetration it achieves, the percentage of all American households connected to cable installations, and accordingly the absolute size of the audience it has managed to collect. The problem is that of the chicken and the egg, in a particularly perplexing form.

While the system remains small, it will continue to grow relatively slowly—perhaps even the current rate of 22 percent annually cannot be maintained. But at some point—perhaps when it reaches 20 percent penetration (although that figure should be construed as nothing better than a crude guess) and consequently can reach a total audience of some 36 million persons in 12 million households—it will begin to possess the economic base that will enable it to provide for the system, particularly if interconnected, totally new programs and totally new services. Beginning at that moment, its attraction will be enormously enhanced, and its rate of growth can be expected to leap precipitously. But how that moment is defined, and when it will arrive, the Commission cannot confidently predict.

We have chosen to assume instead that some new signals, derived primarily from over-the-air production, will be made immediately available, and that growth will continue at more or less the current rate. Upon those broad assumptions, the Commission estimates that by the end of the decade, or perhaps shortly thereafter, penetration will be in the general range of

40 to 60 percent, and that in metropolitan areas penetration is likely to be substantially higher.[1]

The Commission is not wholly at ease with this assumption. It is conceivable, on the one hand, that the point of abrupt growth mentioned above will not be reached during the decade, in which case the lower limit of 40 percent is too high. It is conceivable, on the other hand, that the point of discontinuity will be reached early in the decade, in which case the higher limit of 60 percent is too low. But some assumption must be made, and the range of 40 to 60 percent appears to the Commision to be both conservative and achievable.

3. *Nature of the system.* Systems now being installed are what is called trunk-and-branch systems: signals are in all instances originated at the head of the system, pass along trunks, and are moved into households by cable branching off the trunk lines. It is important to recognize the distinction between this and the telephone system. In the latter, a distinct wire is carried from the head-end to the home; at the head-end it can be connected directly to another wire leading into any other home in that area, or by interconnection between head-ends to virtually any home in the world that has subscribed to the service. The switching process is the most expensive single element in the telephone system; it would be at present prohibitively expensive for the information-laden signal carried by conventional television, and indeed is inordinately expensive for the more rudimentary signal proposed for Picture-Phone service by A.T.&T. Thus the Commission expects that cable television will continue through the decade to be a trunk-and-branch system, with enough elaboration to permit a single area to enjoy the services of several head-ends.

The trunk-and-branch configuration, however, does not pre-

[1] The Commission has relied heavily on a paper prepared under the aegis of the Brookings Institution by John J. McGowan, Roger G. Noll, and Merton J. Peck. An abridgement of that paper will be found in Appendix B; the complete and unabridged paper is available from the Alfred P. Sloan Foundation.

clude a return signal of some sort along the same cable from the household back to the head-end; switching problems are encountered only when it is desired to move such a signal beyond the head-end to a selected destination. Such a return signal might, in principle, assume one or all of three forms; it might be a video signal, such as that received on a television set; an audio signal, such as that received by telephone or on a radio set; or a digital signal, such as those handled by computers. Of these the most easily accomplished is the digital return signal.

By "digital return" is meant, in effect, the capacity to answer "yes" or "no" to any question that may be posed. As anyone who has every played "Twenty Questions" is aware, the problem then becomes merely one of posing the proper questions; any desired amount of information can then be transmitted by a series of "yesses" and "nos", although the process may take some time. At its simplest, digital return can take the form of a device within the receiving assembly which will, upon the depression of a button, send a simple signal representing the word "yes" back to the head-end. On that basis, unending elaborations can be constructed: more buttons, more complex coding of the return signal, more intricately designed equipment.

A second important family of uses for a digital return capacity is represented by those which involve identifying the set to the head-end. A simple circuit can be devised which can be interrogated from the head-end, and which will respond by advising the head-end of a code-number assigned to that set within the system. As will be seen below, the existence of such capacity has significant bearing upon the adaptability of a cable system to pay television, and indeed upon its adaptability for all uses that might involve blocking signals to a given receiver.

As noted above, digital return signals are automatically such as can be received and processed by computers, making it possible to interrogate every set within a given system in a space of time measured in thousands of a second. A return audio signal, although not substantially more difficult to manage at the set than a return digital signal, is far more difficult to cope with at the head-end; it must ultimately be transmitted to human ears rather than to the impersonal fast-moving computer. A return video signal, finally, requires elaborate equipment at the household, enormous channel capacity, and even more enormous receiving capacity at the head-end.

The Commission concludes that for the foreseeable future the digital return signal will be the return signal of choice, and that by the end of the decade it will be a conventional component in most or all cable television installations; indeed, we have so recommended.

4. *Interconnection.* With the existing penetration of cable television, interconnection among installations is of little interest, for the total interconnected system would not be large enough to provide an economic base for the production of special programming; meanwhile the system enjoys many of the benefits of interconnection by being parasitic on conventional network interconnection.

As the system grows in size, however, it will become financially rewarding to effect interconnection within the total cable system. Such interconnection is possible today—and indeed some of it now takes place—by means of micro-wave or cable relay systems. At present such interconnection must be leased from A.T.&T., but recent regulatory decisions by the FCC now make it possible for a major television operator, franchised in many separated locations, to operate his own micro-wave relay link or to rent one from enterprises other than A.T.&T.

What is more interesting, however, is the likelihood in the near future of satellite interconnection. With such an interconnection, television signals would be radiated into space from any one of dozens of earth stations, then radiated back to earth stations capable of covering the entire continental United States and of distributing the signals at low cost to individual cable system head-ends. The Commission is advised that in the immediate future such a system will increase the availability of interconnection and reduce its cost by 50 percent and in the long term by even more.

The Commission concludes that low cost interconnection by means of satellite will be available to cable television by the end of the decade, and that cable television will thereby be capable of assimilation into a national network, or regional networks, on a scale substantially greater than that of conventional television at the present time. (The same, of course, would then be true of conventional television.)

To sum up, the Commission concludes that by the end of the decade, or perhaps shortly thereafter, the cable television system viewed as a whole will be at least twenty channels over all its range and forty channels over much of its range; that it will reach into 40 to 60 percent of all American households; that it will provide digital return signals to computers at each head-end and at little extra expense to other computers at a limited number of selected locations; and that it will be capable of full interconnection at moderate cost.

Significance of Abundant Television

Contemplating now the full image of the cable system that may be reasonably expected to develop during the next decade, one becomes aware that cable television can be something considerably more than a mere extension and expansion of conventional television. The capacity of the system that comes with additional channels, its ability to serve small areas, and to devote channels to particular uses, the expansibility of the system when it is freed from the constraints of the radiated signal, the responsiveness and flexibility that accompany the digital return signal, all combine to make of it an entirely new communications complex. It must be encountered on its own terms, and not on the terms of those communications complexes that preceded it.

Scarcity has imposed, in the past, a series of imperatives on conventional television. In their totality they have enforced upon conventional television, for economic reasons, for political reasons, and for social reasons, the obligation to satisfy those needs which are perceived by the largest number of people at any given moment. Thus television has provided mass entertainment; it has provided national news and regional news, the latter most commonly with emphasis on the spectacular, the former with emphasis on the very broad and the very general. It has necessarily eschewed the particular, whether it be the particular taste in entertainment or the particular need for information. It has been, in general, a vehicle for personal expression only when that expression is generally acceptable. It has established one-way communication between the producer and the consumer, at least in terms of the broad marketplace, but it performs little else in the way of services. None of this is intended, *per se,* to be critical of conventional television: it has performed as it has been obliged to perform, and on the whole done so with surpassing efficiency and skill.

Television of abundance is not merely an augmented television of scarcity. A whole new range of possibilities suddenly appears. The analogy is not to conventional television, but to the printing press. Cable television cannot do what the press does; the printed word continues to be a premium of its own, unchallengeable in what it does well and inimitable in its capacity to work on the higher reaches of the mind. But in important respects—perhaps the most important—television of abundance, unlike television of scarcity, can begin to operate as the press has operated.

Like the press, it can be directed toward a wide variety of uses. The press deals largely with entertainment, in the form of books, of magazines, and to a large degree of newspapers. In those same forms, it is the basic medium of information. It is a vehicle by means of which education, formal and informal, is conducted. It carries the burden of offering an outlet for opin-

ion, warranted or unwarranted, popular or unpopular. It is an
essential intermediary in the provision of almost every kind of
service, private and public. It is irreplaceable in the political
process.

The press accomplished all these ends simply because it is
available in abundance. The publication of one book, or magazine, or newspaper, does not preclude the publication of another; there is plenty of press time available, and in any case
new presses can be built. It is exactly this kind of abundance
that the cable carries over into television, and that makes it
possible, at least in theory, for cable to provide, or very nearly,
the same range of products that the press provides.

Along another dimension, the press can direct itself to a wide
variety of audiences. It can provide, for example, the mass magazine or the best-selling novel, purposefully directed at the widest
possible audience: *Life, Reader's Digest, The Valley of the Dolls.*
It can provide a quite different kind of national magazine or
book, directed toward special audiences of one kind or another:
Foreign Affairs, The New Yorker, Fanny Farmer's Cookbook. It
can provide regional publications or local publications; the van-
ity press will even publish a book directed at none but the author
and his intimate friends. Again, this comes about simply because
one use of the press does not preclude another. The television of
abundance has the same characteristic.

The press can be occasional. For their own good reasons,
magazines come out at regular intervals, and publishers prefer
to have their spring lists and their fall lists diversified and filled.
But the individual who wishes to publish a pamphlet can do so,
or the special occasion that warrants a special commemorative
booklet can readily be obliged. The advertiser can buy a direct-
mail campaign whenever he chooses to spend the money. Once
more, it is the copiousness of the press that makes all this possible—a copiousness that cable television can share.

Finally, the press is not exclusively reliant upon either the advertiser, the reader, or those who wish to purchase its services. A magazine may choose to derive its revenues from those who advertise in its pages, or those who are willing to purchase the magazine, or (in most cases) some kind of combination of the two. The club that wishes to send out notices of its next meeting pays for the service and makes no further demands. Here, too, cable television can enjoy, in principle at least, exactly the same flexibility.

The analogy cannot be pushed too far; like all analogies, it has its limits. The differences between the press and cable television are considerable. The most obvious is the difference in cost of production. A few dollars at a job-printing shop will buy a thousand flyers; there is very little in television that a few dollars will buy. To put the most simple message on television involves the collaborative efforts of a great many people, and does not come cheap. There is, in short, inevitably a cut-off point for cable television that must reduce, in some degree, its flexibility and its real copiousness, as distinct from what it might possess in theory.

There is furthermore a need for special kinds of talent which cannot perhaps be completely met. That same power which is the pride of television requires in some degree a commensurate power on the part of those who serve it. A bad book may be skimmed for whatever it may chance to contain; a bad television presentation can be intolerable.

There is, finally, a profound difference between a product that persists and one that is as evanescent as a television presentation. The television production makes a single appeal (or if it is repeated a few times, a few appeals) to the entire potential audience, and in immediate competition with other appeals. A book or magazine is available over an extended period of time to all its audience. Because a television presentation must be so intensely competitive for its audience, it must be more elaborate than a corresponding print presentation, and must surround itself with promotional devices that will attract audience; these two are essential elements in the cost and the effectiveness of even a copious television system.

Those are a handful of caveats, but the most important caveat remains to be stated. The printing press exists; it has been in existence for five centuries. It has demonstrated its utility, and won its acceptance. Television of abundance is at this moment a concept and not a reality. It may not happen. It may be regulated out of existence, it may prove to be nothing that society really requires or desires, it may founder on the shoals of economic appeal.

But at the least, it should be examined, as far as it can be examined. It is to that examination that this Report now turns.

CHAPTER FIVE

CABLE IN TRANSITION

The cable system that this Report has postulated promises a television of abundance, very nearly as copious as the press, providing all that has come to be expected of conventional television and an endless range of new services to the home, to the institutions society has erected to serve its needs, and directly to the public itself. But to postulate such a system is not to bring it into existence, and the system does not now exist. As it stands today, cable television serves 9 percent of all households rather than 40 to 60 percent; its channel capacity, in most instances, is far less than twenty; it is not worth enough to the advertiser or the entrepreneur to warrant interconnection on any sizable scale. To begin to fulfill any of its potential, it must grow.

In principle, forced growth could be imposed upon it by Congress and its agencies. The electromagnetic spectrum is regulated by the government on behalf of the public: a large portion of the spectrum has been assigned, by the government, to television, and could be withdrawn by the government in a phased transition from conventional to cable television. As we have asserted, technology argues for such a course. But there is no

47

present temper for it, and consequently no likelihood that any such course will be adopted. Instead, cable television must grow, if it is to grow at all, by its own efforts, and it is perhaps not too much to request that government take no extraordinary steps to hinder it.

On the one hand, then, there is the system in being; on the other hand, the mature system this Report has postulated. Between there must be some intermediate stage during which the system grows to size, and during which the cable operator must constantly seek out means by which he may continue to attract new subscribers and retain old ones. To do so, he must offer the subscriber everything that conventional television offers, and something more besides. Each half of that proposition presents its problems and its ambiguities.

Cable's Access to Programming

To offer, in any given area, what conventional television already provides, the operator must be able to retransmit the broadcasts of local stations. All cable franchises do just that, but it is not entirely clear that they possess, under present copyright legislation, any such right. One Supreme Court decision has given the cable operator limited rights, but the breadth of that decision has yet to be tested.

As for "something more besides," the complexities are abundant. In New York, and perhaps in Los Angeles and Chicago, where the economic base is broad and entertainment resources plentiful, he may be able to offer limited first-run entertainment of his own. But for the most part, first-run entertainment will be beyond the reach of his purse as long as his installation, and the cable system as a whole, are young or adolescent. He must fall back upon second-run and subsequent-run material: "I Love Lucy" the third or fourth or fifth time around—the fare with which independent stations have made most viewers familiar.

To procure such programming he must force entry in one way or another into the well-established relationship that has come into being between broadcasters and the owners of the copyrighted material that they transmit over the air—a relationship that reaches the ultimate in intimacy when the same organization is at once broadcaster and copyright owner.

He cannot expect a warm welcome from either party. The broadcaster, if he sees in cable television a threat to the value of his own enterprise, may well decide that his own best interest lies in strangling cable television at its birth. The copyright owner's reasoning may be somewhat more intricate. On the one hand, the proliferation of cable systems seems to extend his market and consequently his opportunity for sales and profits. But he may also reason that the number of people who are willing to see a program twice, added to those willing to see it a third and fourth and fifth time, constitutes an absolute limit on his revenues, and his self-interest lies in limiting himself to the outlets with which he customarily deals (and with which, in addition, he enjoys a comfortable relationship) rather than in seeking out new outlets at the expense of the old.

Seeking to force entry into the market, in one way or another, the cable operator is bound by rules set by the FCC regulating what he may and may not retransmit, and by copyright laws, established by Congress and interpreted by the courts governing the terms under which he may transmit. Neither the regulations nor the laws offer a firm guide. The first change regularly; they are in another process of change at this time. The copyright laws, admittedly outdated, are under great strain; their necessary revision is being used as a battleground by cable and broadcast interests.

At present, cable operators carry on in a jungle not entirely of their own creation. They rebroadcast all local signals, although it is conceivable that some day they may be called upon to make substantial back-payments for copyright violation. They bring in network signals from over the horizon, where their own areas are deficient in network coverage, again under the threat of court-supported demands for back-payment. They bring in distant independent signals, by means of micro-wave links, with consequent damage, in some instances, to the local broadcaster. And they are beginning to make their own arrangements, wherever they can and wherever it appears profitable to both sides, with copyright owners.

The networks, as networks (rather than as owners each of five valuable VHF stations), stand somewhat outside the battle and are yet intimately involved in it. As networks, they seek the most complete possible coverage and are in principle indifferent to the manner in which their signals are transmitted. But dependent as they are today, and are likely to be for some time to come, upon the adherence of their affiliated stations, they respond to the fears of those affiliates that cable installations by fractionating audience will reduce broadcasting income. And as themselves profitable enterprises, they are understandably dubious about a technological development that may well, in the long term, upset the order of things.

Of all those who are involved, the point of view of the set owner is most easily stated: he would like more diversity of program choice. Squabbles within the industry interest him not at all.

Grappling with all these complexities, this Commission has shaped its recommendations on the basis of broad principles that emerged from its considerations. (1) The interest of the viewing public in a diversity of program choice is paramount. If that diversity can be produced equitably, regulations and legislation should be shaped in a manner that will provide it. (2) In any case, there will be such diversity only if the creative process is fostered and rewarded. A healthy program production industry is essential to the television industry as a whole, however the signal is distributed. (3) As far ahead as this Commission can clearly see, there is no destructive conflict between conventional network operations and cable operations such as to necessitate explicit actions to protect the one against the other. (4) Competition between the local broadcasting station and the cable installation should be managed on the basis of equal opportunity of access to programming.

The recommendations of the Commission deal with cable access to second-run and subsequent-run material; cable rebroadcast of network signals; and cable rebroadcast of local independent signals. The recommendations are based on the broad principles asserted above.

Copyright and Exclusivity

Cable companies now rely primarily on distant signals, imported from broadcast stations over the television horizon, to attract new subscribers. We deal here first with the case in which the imported station is an independent rather than a network station, for it is in that instance that the problem of the distant signal is most virulent.

When cable companies now import distant signals from independent stations, they do not compensate either owners of the stations whose signals they import or holders of copyright on the materials those stations show. The law is not clear on their right to behave so. The only Supreme Court decision bearing on the question was in *Fortnightly Corp. vs. United Artists, Inc.,* in which the court held in 1968 that at least under certain circumstances the retransmission of an out-of-town signal was not a performance within the meaning of the Copyright Act, and hence did not give rise to copyright liability. But the more general question of copyright liability is still in litigation, and in any case may be drastically affected should Congress pass new legislation.

A further problem, closely associated with the resolution of the copyright problem, is that of exclusivity. Almost all television programs that are sold at all to independent stations (or to network stations for broadcast when the network program itself does not command the time slot) are sold on the basis of long-term territorial exclusivity. The contracts preclude the program owner (as for example the producers of a Hollywood film or a network program in its second or subsequent run) from licensing the same show to any other television outlet in the geographical area of the purchaser for a specified length of time, usually between two and seven years. If such exclusive agreements are maintained, the willingness of the cable operator to pay for programming will not serve to meet his needs, for the programs he wants will often not be available.

It is difficult to see why exclusive agreements are as common as they are. They diminish the opportunity of the copyright owner to maximize his returns; they reduce the flexibility of independent stations (and network stations to the degree that they program as independents). They deprive the viewer of a large degree of choice, particularly in time of viewing. Their survival can be laid in part to inertia, and in part perhaps to the bargaining power of network affiliates and a few powerful local independent VHF stations. (They constitute, it might be added, one more burden under which UHF stations are obliged to labor.)

The Commission has asserted its belief that a healthy program production industry is essential to the television industry as a whole. In accordance with that belief, we hold that program producers are entitled to a return from performances of their product, and that cable operators who import signals from independent stations or who wish access to signals by other methods, should be expected to pay fees to the copyright owners. Those fees would be negotiated; there are statutory requirements outside the copyright laws that could be called upon to assure that they are fairly negotiated. We recommend further that exclusive rights within a geographical area be severely limited in time, by legislative provisions, with somewhat greater latitude permitted for first-run programs.

The effect would be to provide for cable operators access on equitable terms to the programming they require to foster the growth of their operations; it would ensure returns on fair terms to copyright owners; it would in consequence provide for the viewing public the opportunity to enjoy, by means of subscription to cable television, greater diversity of choice than is now available to it. For the operator himself, access would be unlimited (except for first-run programs)—he could bargain for as many distant signals or as many discrete programs as his purse allowed and his channel space had room for.

The entrepreneurial consequences of such legislation would, at the outset at least, be highly varied. As noted above, some cable operators and lessees of channels would merely wish to take programs and advertising intact from distant stations, making payment to the copyright owner either directly or via the station and perhaps paying the station itself for its labors in arranging the package. Others will seek subsequent-run programming tailored to their own community's special demands, in which case they will have access to the entire repertory of second-run programs. Under such circumstances, some channels will appear much like today's independent stations, presenting a diet of decade-old situation comedies, westerns, and moving pictures; others might specialize in replays of sporting events or documentary programs. Undoubtedly, specialists will appear who know the available repertory, are sensitive to the needs of special audiences and the sizes of those audiences, and can arrange means of financing distribution; they will provide the basis for wholly new kinds of networks, similar in spirit to the Hughes Sports Network on conventional television.

Under certain circumstances, arrangements similar to those of the American Society of Composers, Authors and Publishers —better known as ASCAP—may take shape, in which program producers in effect pool their product and charge flat fees, graded according to audience reach, for those who wish to use it. In other instances advertisers, having obtained rights to copyright material, will offer it free to cable distributors as an entertainment-advertising package.

Established distributors, in command of desirable product, will inevitably devise inventive ways of denying programming to such small entrants as cable installations and UHF stations as a means of retaining existing profitable arrangements. Effective enforcement of antitrust laws would minimize the damage from such endeavors.

The Commission believes it to be in the public interest to encourage the growth of cable television, and believes that improving access to programs will accomplish that end; it sees neither a necessity nor a warrant for attempting to achieve the same end by relieving cable television of copyright obligations.

Reference is made, from time to time in the paragraphs above, to channel lessees. The Commission has assumed throughout that the procurement of programming will at times be undertaken by the owner of the cable installation, and at times by entrepreneurs who choose to lease all or part of a channel and embark in business on their own. We believe that diversity will best be served if those alternatives are retained. An owner who chose to take command of all the channels in his installation, aside from those otherwise mandated, would defeat at least some of the goals of diversity that cable appears to bring within reach. We recommend, therefore, that regulations impose upon the cable operator the obligation to retain a proportion of his channel capacity for lease; the details of that recommendation will appear subsequently, after the full nature of the system recommended in this Report has been asserted.

The FCC's Position and Our Own

These recommendations differ in certain central respects from those proposed to Congress in August 1971 by the Federal Communications Commission. As this Report has stated, the

FCC has in the past been disposed to favor, in this turmoil of conflicting interests, the broadcasting stations. Its rules have had the effect of barring since 1966 the importation of distant signals into the top hundred markets, and it has been upon those signals that cable television has primarily relied. The effect of FCC regulations has therefore been to stifle the growth of cable television in those areas where most people live.

The rules most recently proposed, however, although they do not and indeed cannot resolve the questions of copyright liability, loosen considerably the regulatory grip. Those proposed rules may be easily summarized. They begin by establishing "minimum service standards" as follows:

1. In television markets 1–50
    three full network stations
    three independent stations

2. In television markets 51–100
    three full network stations
    two independent stations

3. In smaller television markets (below 100)
    three full network stations
    one independent station

In all markets, cable operators are obliged to carry all stations licensed in communities within thirty-five miles. In the top 100 markets, they are permitted to import two distant signals, even if such importation brings service above the appropriate minimum standard; they may import more if more are necessary to meet the minimum service standard. Smaller markets must be content with minimum service. (Separate provisions are made for educational stations and foreign-language stations.)

The proposed rules include also "leapfrogging" conditions. These provide first that in the top 100 markets the first distant signal that may be imported must be a UHF station from within 200 miles, if such a station is available; in the absence of such a station, any VHF station within 200 miles or any UHF station may be imported. Second, in those few markets where a third independent signal may be imported, that signal must, if possible, be in-state or within 200 miles. Beyond those restrictions, importation is unrestricted up to the level of the "minimum service standard" as established above.

The cable operator is not permitted, under the FCC proposed rules, to select the distant station that would most appeal to viewers in his vicinity. In many instances he is obliged to carry a UHF station which, because of the family of disabilities under which such stations labor, is likely to carry weak programming at best.

It is apparent that what the FCC has attempted to establish is a compromise position, which frees cable television to expand in urban centers but limits the amount of competition it can offer. On the whole, and in the present general state of cable television, the compromise favors the cable operator at some expense to the television industry. The program producer is more or less unconsidered: he stands to lose along with the broadcast industry, but he does not immediately gain with the cable industry.

In its own terms, the FCC's solution to the distant signal problem is certainly ingenious. But so long as the copyright problem is unresolved, the proposed FCC regulations perpetuate the state of affairs in which competition in any area between the cable installation and the local independent station or stations is waged on a basis of inequality, and the product of the program-producer is appropriated without return to him.

Payment for Programming Remains the Issue

In any case, we believe that statement of the problem in terms of distant signals, as the FCC seeks to do, obscures the real issue: that of access to subsequent-run entertainment on fair terms by cable franchises.

The FCC's solution is ingenious and praiseworthy, a method of encouraging the moderate growth of cable without injuring the broadcast industry. But the FCC is silent on the issue which we think central: the question of compensation for programming.

If the cable operator is not obliged to pay for programming, then the fundamental inequity in competition between cable and broadcast stations continues. It is simply unfair for one competitive mode to pay for programming while the other one does not. On the other hand, if the cable operator, as we suggest, is required to pay, limiting the programming he can purchase is difficult to understand.

If payment for distant signals is required, distant signals become only one source of the additional programming that cable subscribers want. The distinction disappears between programming rights bought from another station or bought directly from the producer. The system owner should have to negotiate a price with the distant station, a price that takes account both of the revenue to the cable owner in the form of subscriber fees and the revenue to the station in the form of advertising sales. In early years, cable systems will probably have to pay stations. Later, when cable systems are larger, the negotiated price might be zero, or even payment by the foreign station for the additional viewers to whom the cable system provides access.

The difference between our approach and that of the FCC can be extremely significant in the moderately large market area. Instead of binding the cable operator to a weak UHF station within his own state, we seek to permit him to reach out to the powerful independent stations in New York, Los Angeles and Chicago, or perhaps to attractive regional sports programming from an independent station a few hundred miles away. Under our rules, all those choices would be available under a negotiated price.

This arrangement for distant signals is likely to lead to more rational allocation of television programming resources. If the system owner is given the right to import, but must reach a negotiated price with the owners of the material he transmits, he can choose rationally among available options: distant signals, origination by himself, purchase of material not being shown over-the-air, participation in a "cable network." Similarly, creators of programming material will be able to investigate various ways of reaching the television audience, and select the most profitable.

It appears to this Commission that its recommendations with respect to distant signals and copyright provide the opportunity for cable to grow to maturity without government subsidy, or subsidy by those who create television programming. Growth based merely upon access to existing subsequent-run programming is not likely to be rapid, but there is at least some reason to believe that it will feed upon itself; as the system grows upon the basis of what is now available, it begins to gain the capacity to stimulate the production of something more than is now available, and its rate of growth rises accordingly. There are entrepreneurial stirrings that suggest something of the sort is already in progress: the plans of such organizations as Hughes Aircraft Company and General Electric to program specifically for cable television. Even a slow growth of cable television, particularly in the large markets, will stimulate those plans and others like them. Over a decade, as we have earlier asserted, the transition to maturity can be expected to take place.

Carriage of Network Signals

Another form of the distant signal problem arises when the imported station is network-owned or network-affiliated. It is a consequence of the fact that about 5 percent of the nation's viewers cannot now receive the programs of all three conventional networks. We recommend simply that cable operators in such underserved areas be permitted to fill the blanks in network service which were, in a sense, accidents of spectrum allocation.

The network as network—that is, without respect to its owned-and-operated stations and without taking into consideration the special interests of its affiliates—seeks to provide as wide an audience as possible for its commercial sponsors, and makes payments to its affiliates for helping provide that audience. In areas underserved by network affiliates, any given network gains if its own signals must be imported from outside the area; it loses if a competing network is imported.

Since there appears to be no equity in such gains or losses, which arise simply out of the necessities of spectrum allocation, the Commission sees no objection in principle to regulations which permit cable installations to provide, either off-the-air or by importation, the full complement of network signals. This privilege should be extended to those instances in which the local network affiliate chooses not to transmit a network program, as affiliates often do refuse to transmit public affairs and documentaries; in such cases the cable installation should be permitted, if it so chooses, to procure the program by importation of signal.

The Commission recommends that copyright legislation be drawn in a fashion that permits the cable installation to use network programs without copyright fee. We do not believe, however, that the cable installation should be permitted to profit directly, by selling its own advertising, in association with programming which it obtains at no cost from the network; advertising time-slots, if they become available, should be mandatorily allocated to public service announcements.

We recommend also that the current policy of requiring cable operators to carry local stations be maintained and that no copyright liability be imposed for local carriage. No requirement is suggested in this Report that would force the cable operator to provide local public affairs programming or local news. But the interest in localism may be insured by making sure that the local station (which must fulfill FCC license standards for meeting community needs) appears on the system.

We do not think that it is sound policy to force a viewer to choose between the cable programs and local over-the-air service, and if the cable operator does not include the local station in his offerings, it may be impossible for his subscribers to receive it. For the local station, making programs available to cable installations appears to the Commission to be, in a sense, an additional kind of technical standard, under which the station, by virtue of its license, is obliged to make its signal generally and broadly available. To the extent that the cable installation merely rebroadcasts the signal, it is not in direct competition with the station; indeed, if the station is in the UHF band the rebroadcast has the effect of augmenting station audience and hence station returns. As with network rebroadcast, the cable installation should not enjoy the privilege of selling its own advertising in association with rebroadcast programming.

We recommend also that cognizance be taken of the special nature of the Public Broadcasting system. Public Broadcasting is defined as a public service, and supported upon that basis by federal and in many instances state and municipal funds. We recommend that cable operators be obliged by regulation to carry local Public Television stations, and that in the absence of such a local station the cable operator be privileged to import a Public Television signal to make up the deficiency.

The effect of these last recommendations, if adopted, would be to provide for cable systems in the largest television markets the right to offer their subscribers the same service that is available over the air, and in the lesser of the large markets to bring network service up to the accepted level of three networks. Of itself, that would provide little capacity for growth in the very largest markets. In those markets, however, we count upon local origination and new services, together with better color signals, to provide the initial impetus for growth, as they are now doing in New York City and promise to do in other great metropolitan areas.

A *Necessary Distinction*

There is an apparent asymmetry in these recommendations that some may find disturbing. The cable operator is permitted, and indeed obligated, to carry network signals and local independent signals without payment of copyright fee. He is obliged to pay his way to gain access to other programming. But the arguments of equity would appear to apply equally throughout. Since in each instance he receives something of value to himself, he should therefore, it might appear reasonable to argue, in each instance make some kind of payment.

We believe the asymmetry is not all that it may appear to be. In carrying network signals, the cable operator assists the network in gathering the audience from which the network derives its revenues; he is doing at no cost to the network what its affiliates do for a share in the proceeds. Much the same is true when the cable operator carries the local independent station; here too he gathers audience (perhaps substantial audience in the case of a UHF station) at no cost to the station. If the incremental audience, in either case, is large enough, the copyright holder can take it into account when he strikes a bargain with the network or the independent station. Only when the cable operator seeks programming that is directly competitive with local radiated signals does he hope to derive direct benefit from that programming, and when he seeks to accomplish that end by the use of copyrighted materials, he should expect to make payment for that material.

On the whole, the Commission believes that over the long run the outcome of requiring payment in all cases would differ little if at all from the outcome that would follow adoption of these recommendations. Cable television would come to terms with networks and local stations, make or receive its payments, and evolve in its own fashion. But that evolution would be distinctly slower, and the maturity of the cable system far longer in arriving. Because we believe that the reasonably rapid growth of cable television is in the general public interest, and because we believe the short-run losses to copyright owners are surely modest, we have taken the view expressed in this Report. The costs of operating within a more strictly logical system are simply not worth the gains.

In the end, cable must grow as conventional television has grown: on the basis of its own accomplishments. As it takes on an identity of its own, the current debate over distant signals and the passion it arouses, as well as the disputes concerning the rights over local broadcast signals, will come to appear insignificant stages in the growth of a total television system. The recommendations made here for the period of transition are intended primarily to be as equitable as possible to all the parties concerned during the difficult period when that total system is gradually taking shape.

CHAPTER SIX

ENTERTAINMENT ON THE CABLE

Viewers turn to television primarily for entertainment, and entertainment is most of what television provides. It is, furthermore, entertainment of a particular kind, akin to the national fiction magazine (which it has very nearly put out of business), the national illustrated magazine, the best-selling novel and the long-run moving picture. Conventional television provides such entertainment in the form of network and motion-picture studio production transmitted over the national networks, sporting events of wide appeal, and on a lesser scale by means of first-run syndicated material delivered on film or tape to local stations.

The costs of production and distribution are borne directly by advertisers, presumably out of income earned from the purchases of those who watch (as well as those who do not). In some sense and in some degree, then, the viewer pays for his entertainment, although how much he pays is certainly arguable, for it can be maintained, although not very persuasively, that television advertising also produces economies for the advertiser which he passes on to the consumer in the form of price reductions.

But the balance of conventional television is such that for the most part the viewer is obliged to pay (however little) not for what he would particularly wish to see but for what the total audience, or some very substantial part of that audience, would wish to see. The element of choice is distributed over the population at large. The program that might deliver a high degree of satisfaction in 10 million households is displaced in favor of the program that will deliver a moderate degree of satisfaction in 20 million. With limited channel space available, that is not an entirely unreasonable state of affairs; it can be seen as a democratic solution to a problem that must be solved in one way or another. With unlimited channel space, the problem — at least in that form — vanishes, although inevitably certain economic problems appear in its place.

The cable cannot, of course, provide unlimited channel space. Its services to diversity can be accomodated, however, by an installation which is purposefully designed to make room for entrepreneurial agility. And those services may be precluded in any area where the franchise-holder setting himself up in business does so on the basis of minimum channel capacity. The cable operator who believes, in some locality, that he can sell his subscriptions on the basis of three local stations and a few distant signals may be tempted to install a system just capacious enough for those demands, locking his subscribers into place with inferior service. Some of that is now happening: for every cable operator now installing systems that can handle forty channels or more, such as Akron and San Jose, there are reports of systems content to make provision for twelve channels or even fewer.

Diversity and Channel Capacity

Additional channels, over the number that appears immediately useful, can be accommodated inexpensively when the original installation is made, for the major cost of installing a system is the cost of physically laying the cable, and not the cable itself. It is thus no large hardship upon the franchise-holder to oblige him, as a condition of his franchise, to provide a respectably copious installation. As a floor to the provisions of municipal franchises, the Commission recommends that a federal minimum should be established, as a necessity in preparing the way for a national cable service. The Commission's technical advisors suggest that the technology and the economics of installation make twenty channels a reasonable minimum. Municipal franchising authorities which believe they can identify even greater need for channels should, of course, be free to set minimums above that floor.

The second requirement, if diversity is to be accommodated, is that a diversity of imaginations and skills be applied to providing programs. This requirement is not automatically served by the provision of adequate channel space, for there must also be access to those channels. A cable operator who chose to take command of all the channels within his system, aside from those mandated by the municipality for non-entertainment uses, would defeat at least some of the goals of diversity that cable television appears to bring within reach.

We recommend, therefore, that regulations impose upon the cable operator the obligation to offer a proportion of his channel capacity for lease. The nature of entertainment enterprises suggests in general that much of the leasing will be on a long-term basis, so that an entrepreneur may take command of a channel over a period of a year or more and seek to build an audience for his wares. But leasing on shorter terms, perhaps even by the hour, is to be encouraged. The goal of such provisions is a far more widely shared control over television programming, and new and flexible mechanisms for obtaining access to the public.

With leased channels available, the variety of users of cable is limited only by imagination. The local motion picture distributor might lease a channel to make his films available directly in the home; the newspaper a channel or part of a channel to provide a news service. The boxing promoter will look to cable as a substitute for closed-circuit theater distribution. Groups who may be so minded will be able to lease a channel, at appropriate times of the day and week, for high quality children's programming.

One question remains. The costs of programming must be met, and television programming never comes cheaply at the level of polish to which the audience is accustomed. The returns to the producer can be derived from increased advertising rates, justified to the advertiser by the nature of the audience he now deals with and its relation to the product he seeks to sell: life-insurance for the newly-wed, air-travel for the businessman, electrical appliances for the housewife. It can be derived also from direct subscription payment by the viewer himself. It can be derived, finally, from a combination of the two.

Payment by the Viewer

Direct payment itself may take several forms. It may be made by program, in which case the subscriber pays for exactly what he wishes to see. Payment may be made by the channel; that is, a channel otherwise dark on the subscriber's set may be activated upon agreement to pay a surcharge on the monthly subscription fee—a procedure fully analogous to subscribing to a magazine. Or it can be made by both program and channel, perhaps under an arrangement by which subscribers to the pay channel benefit from reduced rates for special events.

The cable system prefigured in this Report will provide with ease the technical means by which pay television can be accomplished. Blocking or unblocking a given channel can be accomplished physically, by insertion of appropriate filters on the feeder line to the residential unit. With digital return, associated with addressing equipment at the set, blocking, unblocking and billing can all be accomplished electronically by remote activation of electronic valves on the drop-line, and by monitoring the entire system to determine channel and program usage.

Almost certainly interconnection will be required to make a pay television system work efficiently, for only if the system is interconnected can the entrepreneur manage the necessary national promotion of his programs. Existing technology makes that interconnection available by conventional means; by the end of the decade interconnection will be available, on much more moderate terms, by satellite. The technology of the needed system, in short, is either immediately at hand or directly upon the horizon. The economics, and the acceptability, of any of the three forms of pay television remain to be determined. In all likelihood, only experimentation and experience will determine them. What the Commission has to offer are in the first instance nothing more than some general considerations concerning the demands of special audiences and how they might be met.

It is clear that pay television provides a means to restore some degree of choice to the individual viewer. It is not total freedom of choice: the viewer who, for reasons of his own, wishes to spend an evening watching and listening to an All-Girl Dixieland Jazz Band cannot do so simply by offering to pay a fee. But viewers can be accumulated with tastes in common of a sort different from the general tastes of the total population, and can be served by cable television: lovers of popular music can enjoy the occasional extraordinary event in the world of popular music, or a channel which provides nothing but popular music and which may even provide, from time to time, the nostalgic delights, whatever they may be, of an All-Girl Dixieland Jazz Band.

To work the process out in detail, one might begin, perhaps somewhat provocatively, by considering opera lovers, who it is to be presumed are somewhat fewer than popular music lovers. Since opera is available only in limited locations, and at extremely high prices, there are no data upon which to estimate the number of opera lovers; let us arbitrarily assume that they are to be found in 1 percent of all residences in the United States—600,000 homes. Those households will be distributed throughout the country.

By the end of the decade it will be within the technical capacity of the cable system we have postulated here, with its satellite interconnection, to deliver a performance of the opera to approximately 300,000 of those homes, either simultaneously or by means of recording and reproduction, and without interfering with the ability of the rest of the population to see and hear whatever else it may be they wish to see and hear. That is the great miracle of the cable system, in its mature state.

With that capacity, the entrepreneur can expect to derive revenues from one or more of three sources. He can, if he chooses, continue to charge admission to the opera in the locale where it is presented. He can require a charge of perhaps $1 (or more, or less) plus channel and distribution costs from each television viewer who chooses to watch. And finally, he can sell advertising on the cable television system to those who may be particularly interested in reaching the kind of audience that is attracted to opera. For the first time, he can hope to make up his deficit and show a profit. (The first outcome of which is likely to be more and better opera.) And for the first time, opera will be available (although in circumstances somewhat less than the ideal circumstances of attendance at the opera itself) for all those who are interested enough to pay a modest price.

A Cultural Channel

Payment by program, however, is not the only recourse. The opera audience is a sub-group of a wider cultural audience, made up of those who are attracted to opera, telegenic presentation of the graphic arts, or to symphonies or other kinds of classical music, or to many or all of those. There are also those who, strictly speaking, lie outside that cultural audience, but who would pay a modest fee to have cultural attractions available to them on those rare occasions when they feel the need for it. As a practical matter, there are also those who feel no need for cultural programming themselves but believe they should have it for the sake of the children, or merely to impress the neighbor.

There is no possible means, at this time, to estimate the total audience for a cultural channel. But one can speculate that if such a channel were available at a cost between $1 and $2 a month, it might conceivably attract as much as 25 percent of the available cable audience. On our assumptions, that would be 7.5 million homes, providing in total revenues $90 million to $180 million a year. With such revenues, first-run and subsequent-run cultural programming should not be excessively difficult to come by, even with channel costs and distribution costs deducted.

The problems of creating such a channel are currently being studied by the Lincoln Center for the Performing Arts in New York—the largest institution of its kind in the world and consequently possessed of the largest operating deficit. It may well be that the ingredients for such a channel will be at hand even before cable television is ready for it.

The cultural channel more or less identifies itself. The problem for the cable television entrepreneur will be to define other audience groups, comparable in size to or larger than the cultural audience, to whom special appeal may be made and for whom special programming may be designed. The audience for news, documentary programs, public affairs and the expression of opinion is one such group that comes immediately to mind, but perhaps falls outside the area of entertainment and involves other problems; it will be dealt with in the next chapter. There is also the audience that is represented in the world of magazines by *The New Yorker* and the like. There are the special audiences of young mothers, of the aged, of teen-agers. One might consider a "professional channel," appealing in alternating time segments to doctors, lawyers, educators, engineers. One might well consider an "ethnic channel" appealing in alternating time segments to the various ethnic sub-groups in the community. Many of those channels would intermix entertainment, news and public affairs programming.

This Commission has neither the desire nor the competence to play entrepreneur. If cable television follows the pattern of other communications activities (and if regulation does not keep the entrepreneur at arms' length) it is reasonable to assume that many such ventures will be attempted, and most will fail; those which meet the needs of the market will succeed and will endure. The Commission concludes only that the field should be laid open for the entrepreneur, so that he may take the risks and reap the reward where reward is to be found.

CHAPTER SEVEN

IMPACT OF CABLE ON OVER-THE-AIR TELEVISION

If cable is permitted to develop, and if the predictions of this Report of its channel capacity and its penetration are reasonable, there will take place a transformation of American television from a system in which virtually every viewer relies upon local over-the-air stations to one in which a great many viewers subscribe to cable. Such a transformation will have vast consequences for the television industry.

To the extent that the development of cable television costs certain businessmen money, or deprives them of profits they would otherwise have earned had government intervention barred or limited the spread of cable, there is no case for public intervention in favor of the established industry. In an environment of fair competition, one of the risks of investment is the appearance of a new and preemptive technology.

But for quite different reasons, the impact of cable on over-the-air television is worrisome, and must be considered carefully. If cable will deprive some citizens of satisfactions and

71

pleasures they now enjoy, such deprivations are an appropriate
subject of public concern.

One must consider first those who live in areas of such low
population density that it will be uneconomical (in the absence
of new techniques) to bring cable to their households. They
number perhaps 15 percent of the population. Certainly the
early stages of development will not destroy over-the-air
television, so far as these viewers are concerned. The cost of
transmission is a minor factor in the production and distribution
of television programs, and for some time to come the advertiser will be willing to pay the small extra sum that is necessary
to continue to reach that part of the audience—a very large
part during the decade—which is not on cable.

But even during that period, there may be local rural stations which simply drop below the level of profitability and
go off the air. In some instances the matter will be accommodated by a simple transfer of license: a cable franchise holder
within the general region, to whom the costs of maintaining
transmission facilities would represent only a small increment
upon his ordinary operating costs, might find it worth his while
to maintain service to remote areas.

But the Commission believes that in any case the government
should act to insure that all potential viewers enjoy some
minimum television service. There are available cheap and
painless alternatives for achieving that goal: it might be dele-
gated to the Corporation for Public Broadcasting, which would
then operate stations in the public behalf where such stations
would be privately uneconomical; it might be achieved by
government subsidy, in the name of public service, to selected
rural stations. Finally, the development of technology might
ultimately make practical, at reasonable cost, a cable installation program similar to the Rural Electrification Program
of the thirties.

The mobile television set, in the automobile or carried to
the beach, is a less significant problem. The first of these does

not exist in any large numbers, and is not likely to become more than a plaything; radio remains the instrument of choice in the automobile. Television in resort areas can perhaps be provided by low-powered transmitters or cable outlets, in the absence of conventional television. If it is not provided at all, the deprivation is truly insignificant.

There are problems as well in densely populated areas. Self interest may lead cable operators to believe that they can maximize returns by "skimming" their franchises, providing services in middle- or upper-class neighborhoods and avoiding construction in poorer neighborhoods. Yet poor citizens rely most heavily on television for entertainment and information, and services especially important in poor neighborhoods may be among the great benefits to be expected from cable. The Commission recommends that the provision of service throughout the franchise area be made, in one form or another, a condition of municipal franchises, with due regard for local needs and local conditions.

Effects on Quality

The considerations above deal with *quantity* of service; the Commission is confident that reasonable quantity will be available over the air so that no portion of the viewing public will become unserved, and that the availability of cable service will not be harmfully restricted by arbitrary decisions. The *quality* of over-the-air service is another question.

Over-the-air television will not be the same service it is today if cable delivers both over-the-air signals and additional programming to a substantial part of the population. As cable systems become an important market for the creativity and skill of the performer, the writer and the producer, certain programming that would otherwise find its way over the air will be bought by cable operators and lessees. Fractionation of audience will mean lower income for some broadcast stations and networks, and might reduce the funds that conventional television can now apply to program production; in particular, expenditures for local news and local entertainment, already low, might decline or even vanish.

The householder who subscribes will enjoy, for his basic subscription fee, more television than he now receives; for additional fees he will have still more television and more diverse television. But the resident who spurns cable and wishes to rely on over-the-air signals is likely to have less, at least in quality, than he now commands. It is true that conventional television is not "free" if the total audience is taken into consideration, but since payment is made indirectly the amount of payment is left, in some degree, to the election of the individual viewer. The loss of that discretionary power is in some measure a social cost, not to be denied or minimized.

Control of "Siphoning"

The issues presented by the potential for flight of programming from free to pay television is most sharply illustrated with respect to the sports audience. It is, during most of the time, a special audience as that term has been used here. But a few times a year, on the occasion of sporting events of extraordinary interest, it becomes a national mass audience. These sporting events, in addition, are different from conventional mass entertainment in a significant way: each of them is unique. "Bonanza" on one network may be paralleled by "The Man from Shiloh" on another, but the Superbowl is the Superbowl: as a distinct program each Superbowl exists only once. The viewer who wishes to see the Superbowl has no alternative viewing.

It is generally conceded that pay television, if it comes into existence on any large scale, will be able to outbid the commercial sponsor for rights to extraordinary sports events. (Indeed, it will also be able to outbid closed circuit television as it was employed to transmit the Ali-Frazier prizefight and as it may be expected to be employed for outstanding sports events in the future.) There is accordingly the imminent threat of what has come to be called "siphoning"—the gradual disappearance of outstanding sporting events from "free" television[1] in favor of pay television.

The FCC has responded to this threat by "anti-siphoning" rules. Under those rules, as recently amended, pay television will be forbidden to make charges for any sporting event of a sort which has been presented over free television during any one or more of the five years immediately preceding the event. The effect of the rule would be to maintain on free television not only such events as the Superbowl, the World Series, basketball and hockey playoffs, both amateur and professional, and the Summer and Winter Olympics, but also most major league baseball, and football, and a whole range of other sports.

Those who favor anti-siphoning rules argue that something very like a vested public interest exists in those sports which have for years been transmitted over conventional television. Sports are conducted as business enterprises, but there is a general feeling that the product in which they trade is something more than another material product. Under the assumptions of this Report, even by the end of the decade some 40 to 60 percent of the total viewing audience would be unable to view these unique sporting events, even if they were willing to pay for them, for they would remain without cable television. Those on the cable would be obliged to pay, and perhaps pay high prices, for something which they had become accustomed to believe they had purchased with their television sets.

Against that view, it is argued that in practice the net effect of anti-siphoning rules will be to diminish, in time, the sports that would otherwise be available to the television audience,

---

[1] In this context, "free" television is taken to include cable television for which only the basic subscription fee is charged.

both on the cable and those who remain restricted to conventional television. This will occur, it is maintained, in part because a viable sports pay channel would offer less widely popular sports as well as those which are widely popular; in part because in the absence of pay television economic pressure will ultimately move the unique and highly popular events from free television to closed circuit theater television, over which the government has not asserted regulatory powers. (Gross revenues from the Ali-Frazier fight were reported to be of the order of $20 million. Even after the high costs of distribution and theater facilities are deducted, no commercial sponsor can compete for such an attraction. Pay television in a mature cable system might surpass it many times over.)

Argument and counter-argument are, in this instance, nicely balanced. It appears to this Commission that some anti-siphoning restrictions are appropriate, but we are convinced also that any anti-siphoning rule is more appropriately left in the hands of Congress than of the FCC; tailoring appropriate limited restrictions will be a sensitive task with harsh financial overtones for the players and owners of certain athletic teams. We would favor an outcome in which certain playoff and championship events, specified in the legislation itself, would be reserved for conventional television so that they might be available to all television viewers.

A second particular kind of special audience must be more loosely defined: it consists of those who are willing to pay some kind of fee for essentially the same programming that is available on conventional television. They may be led to pay that fee for access to the performances of unique talents—Elizabeth Taylor, Bob Hope, Dean Martin—or for programs free of commercial interruptions. The outcome here too might be a kind of siphoning, in which highly popular programs move from free to pay television.

No reasonable analogy with sporting events can be maintained. The unique performer on television is, more often than not, a product of television itself; he can be and usually is tied down by contractual arrangements during the peak period of his value. Where no such contractual arrangements exist, there is in any case no vested public interest in his appearance on television, for he has been traditionally free to offer his services elsewhere, in Hollywood, on Broadway, or on the night-club circuit.

Nor is the Commission moved by the argument that popular programs might move from one medium to the other. In the long run, both free and pay television will have the same opportunity to create popular programs and to bind performers to those programs for extended periods of time. Meanwhile, equity requires that the performer be free to sell his services, under contract or not as he may be able to negotiate, where they will provide for him the most satisfactory return.

The problem of siphoning is ordinarily taken to refer to the passage of popular programming, whether in entertainment or in sports, from conventional television to pay television. But there is also reason for concern over a quite contrary form of siphoning in which the less popular programs are involved.

The great predominance of programs that appear on the three networks are directed toward the mass audience, and are measured in terms of the size of the audience they enjoy. But commercial television does from time to time present serious drama, classical music and skillfully produced documentary programs. The very fact that they are presented on channels normally given over to mass entertainment means that they attract far greater audiences than they would attract on channels directed toward special audiences. There is a risk that with the development of a thriving pay television system, such programs will move to pay television, leaving those dependent on over-the-air television with access to mass programming and mass programming alone.

Any outcome of that sort would be undesirable, but it does not appear likely during the period covered by this Report.

Public Television might serve to make such programs available over the air, but they would never attract on Public Television audiences of the size they attract at present, merely by reason of habitual allegiance to network stations. Those who program for the mass audience have, out of their own sense of responsibility, always found time for considerations other than those of audience ratings. We expect this would continue.

Impact on the Networks

It is unnecessary to argue that the existence of popular entertainment on pay television will have the effect of fractionating audiences, and of reducing the audiences now enjoyed by the major networks. This appears to the Commission to be the kind of risk that technological development imposes upon any investment in an environment of fair competition. A more serious, but more conjectural, danger to conventional television arises from the possible future change in the operation of the three major networks.

What cable television may provide is competitive opportunities for those who wish to enter the field against the three established networks. So far as conventional television is concerned, the FCC table of allocations is such that national coverage, by means of affiliation agreements, is available to only three networks; indeed, it does not quite extend to the third. Such a state of affairs makes it difficult for the entrepreneur whose goal it might be to supplant an existing network with one of his own. A cable system with a high degree of penetration might make the odds on such an untertaking markedly more favorable; beginning with the sizable economic base of a cable network, the entrepreneur might then bid against the three existing networks for affiliation agreements and ultimately capture access to the mass national audience. Since there is some evidence that mass advertising is sufficient to support only three networks, on the present scale, the likelihood is that the consequence would be the replacement of an existing network rather than the creation of a fourth network, even if cable penetration should extend farther than the Commission expects within the decade.

There are, however, further questions about the stability of the present network system under the impact of cable. They arise out of the relationship between the television network and its affiliate. The affiliation agreement requires the network to pay a fee to its affiliate; in effect it purchases the audience the affiliate is able to provide. On the basis of the audience it accumulates by purchase from all the affiliates (as well as those of its owned-and-operated stations, limited by FCC regulations to five on the VHF band) it sells the advertising which provides the network with its revenues.

If, however, cable installations in any marketing area grow beyond a certain size, it becomes possible for the network to consider buying its audience directly, rather than through an affiliate, by leasing channels on the cable installations that cover the market. Because the size of its audience, within that area, would necessarily decrease to the extent that the network has lost coverage in the area, it would suffer a diminution of income from national advertising. To compensate, however, it would be freed of the necessity to pay the affiliation fee, and what is even more important would gain access to the high-profit local advertising that the affiliate is privileged to insert within the interstices of network programs.

Obviously, at some point of penetration short of 100 percent, the lines cross, and the gains outweigh the losses. At that point, all other things being equal, it becomes advantageous for the network to cut its affiliation ties and go into business for itself. The loss is suffered by the affiliate, which now is forced to become an independent station and is deprived not only of network revenues but of the programming upon which depends its ability to attract and maintain its audience. There is, of course, a loss to the public as well: those outside the reach of the cable system, or who do not choose to pay the cable fee, would be deprived of that network's program.

In national terms, there seems to be little risk of any such outcome during the period of time covered by this Report: it is reasonably certain that nothing of the sort is likely to happen at penetration of 40 to 60 percent. Moreover, even in the likelihood that certain local franchise areas will achieve penetration of far more than 60 percent, perhaps ranging as high as 90 percent, such penetrations are not likely to affect entire marketing areas, and it is with the marketing areas (or the signal area) that the network is concerned, not the small enclave within it. Nonetheless, the possibility of affiliate shedding should be a matter of concern, as it approaches and perhaps even occurs here and there. If it occurs on anything beyond the smallest scale it will open up questions of regulatory policy concerning network leasing of cable channels, which would have to deal with the preservation of over-the-air television for those for whom it is the only recourse, or who might prefer to make it so.

On balance, then, the Commission sees no need, during the period with which it deals, for regulatory intervention in the relationships between the established networks and cable installations, at least so far as network access to cable channels is concerned. It concludes only that at some point, perhaps arriving toward the end of the decade, it may be necessary to reexamine the question in the light of the experience that has been accumulated.

There is one final area where the danger to viewers of conventional television is great: news and public opinion. Because of the importance of the subject, we devote the next chapter to the contours of the effect of pay and cable on the dissemination of news.

In our consideration of all these issues, we find ourselves in clear conflict with the FCC. In its report to Congress, the FCC has written ". . . our objective throughout has been to find a way of opening up cable's potential to serve the public without at the same time undermining the foundation of the existing over-the-air broadcast structure." This Commission does not feel the weight of any such objective. We have sought to maximize the service to the public that can be provided by television as a whole. If, in that process of maximization, the existing structure of over-the-air television is undermined, we believe that the public interest must still remain paramount.

We do not believe that in the short term the structure of over-the-air television is seriously undermined; we believe further that in the long term over-the-air television can adjust to the developing new situation. But in any case, if over-the-air television is to fall victim, in some degree or another, to technological change, it is in no different position from any other enterprise in which investments have been made, and possesses no greater right than other industries to protection from technological change. It does not appear to the Commission that the industry needs or warrants further protection by regulatory agencies.

CHAPTER EIGHT

NEWS AND OPINION
ON THE CABLE

One threat of cable television is that it may reduce the diversity of network hard news programming and affect seriously the availability of network documentaries. One promise of cable television is that it is capable of replacing network diversity and amplitude with a diversity and amplitude all its own that will produce an even richer system of television news.

Television, in terms of the allocation of its time, is primarily a medium of entertainment. But it is also the source to which most Americans turn for news. Beyond the hard news, television carries public affairs and documentary programming, although it begrudges them prime time. Television, moreover, is a powerful disseminator of opinion: it appears in the form of commentary included within news programs and explicitly identified, editorials set outside the news programs and presumed to express the opinions of those who control the station, and statements by participants in public affairs and documentary programs.

In a less direct manner, commercial announcements may be construed to carry opinions beyond those conveyed in the claim that one product is better than another of the same kind. Two opinions come at once to mind, toward both of which the FCC has directed its attention: that cigarette smoking provides satisfactions that outweigh its dangers, and that the construction of a pipeline across Alaska is in the general public interest. Finally, network television is frequently accused of conveying opinion covertly on its hard news programs in the form of biased reporting or editing, or even by means as subtle as the twitch in one Cronkite eyebrow.

Leaving opinion aside for the moment, the position of news and associated programming on the networks is somewhat anomalous. Network news is an extremely expensive enterprise. For the most ambitious of the three networks, costs probably amount to something in the neighborhood of $50 million a year; revenues, at least recently, have been within a few percent on one side or the other of that figure.

Over most of television's history, such programming has been presented as an out-of-pocket loss to the networks. More recently it is probable that the news operations of at least two of the three networks have been conducted at a profit, although perhaps not quite the profit that networks would prefer. Documentaries and public affairs are presented at a substantial loss, which is recovered by the extremely successful early-evening hard news programs. Documentary and public affairs programs also diminish audience not only in their own time periods but in adjacent periods, and are broadcast only with great reluctance by many network affiliates.

Yet all three networks devote great time and energy to their news broadcasts, even though at best they are only marginally profitable, and produce far more documentary and public affairs programs than their direct economic return would justify. The non-economic reasons for such generosity are manifold. Perhaps they have to do in some small part with the egos of those who own and manage stations—egos that would not be satisfied by merely presiding over mass entertainment.

They have to do, in larger part, with the necessity to create an image of respectability for the network that will attract the commercial sponsor, and in particular the large corporate sponsor—magazines and newspapers feel the same pressure. They have to do with the maintenance of good relations with the FCC. Strictly speaking, the FCC lacks regulatory authority over the networks (although it possesses authority over their highly profitable owned-and-operated stations). Nonetheless, the networks are heated and cooled by winds that blow through the FCC.

Local news and news-associated programming come into being under much the same set of pressures. The local station is also affected by the egos of its owners and managers, and also feels the obligations to create some kind of image of respectability for itself. Unlike the network, the local station is required by FCC regulations to provide local public services, primarily by way of its news department. In principle, it can lose its license if it is nonfeasant or malfeasant; in practice, the FCC issues those rules more vigorously than it enforces them.

What is most burdensome for the local station is the sheer cost of producing local news and documentaries. (The "talking faces" of public affairs programming come relatively cheap.) Even minimum coverage is costly when a full crew and a battery of expensive equipment must be dedicated to half a day's work to capture two or three minutes of film or tape for one-time transmission. The consequence is that local news coverage, except in a few very large cities, is likely to be dismally inadequate. Local stations tend to cover fires (one can always be sure there will be a major fire or two) with one crew, and important set-pieces—a rally at City Hall announced well in advance— with that crew or another. Many stations give up entirely the struggle to produce local news, and content themselves with the minimum of public affairs programming that will keep the FCC mollified.

All this is prelude to a consideration of the effects of substantial cable penetration on the access of the general public to news and news-associated programming by way of television.

Payment by the Viewer

Should the time come when the fractionization of audience brought about by cable television begins to affect network revenues, news and news-related programming will be the first to suffer. The costs of entertainment programming are elastic, for talent costs rise and fall with revenues, and talent costs are the largest component of entertainment costs. (A network program which moves from prime-time to daytime, or from network presentation to syndication, is produced at smaller expense even though the talent may be the same.) The fundamental costs of news programming, however, are the logistical costs and the basic labor costs; they can be reduced only by reducing coverage.

One can assuredly conceive of economies. Should only one correspondent and one crew, serving all three networks, be dispatched by charter plane to Peru when an earthquake strikes, each network would reduce its costs for that story by perhaps half. At the extreme, all but the most vital coverage could be pooled, and the networks differentiated (as for the most part newspapers are differentiated) by the manner in which the news program is made up and by the elements which are added in the studio. Such economies, through this decade at least, might enable the network to continue to provide national and international news at much their present level, if in somewhat different form. (They might also, of course, bring about anti-trust requirements concerning access.)

There are, however, no comparable economies for documentaries. At present their cost is diminished at the network level because they employ staff and equipment which must be on hand for coverage of hard news, and are not otherwise fully utilized. Pooling arrangements for hard news would make the documentary, unit by unit, more expensive in a period when the revenues it might attract—even now inadequate—are diminishing.

The promise of cable television lies once again in its capacity to serve, for a fee, the special audience that is particularly interested in access to news and documentaries, or the special audience that would be willing to pay for the convenience of news service provided more or less upon demand. If there can be defined a special audience in either of these senses akin to the cultural audience defined in a previous chapter, one can perform simple arithmetic upon the economics of pay television news.

On the assumption that each of the three network news services operates at a level of $50 million a year, and that each serves, over the course of the year, one-third the viewing audience or 20 million households, it is clear that each network must derive $2.50 per household in revenue for a break-even operation. These revenues come in the form of commercial fees, of which approximately half are derived from the five minutes of commercial announcements transmitted each weekday evening on the early-evening news broadcast.

Against these figures, we may now consider two distinct configurations of cable network news services, each of them charging a fee of $1 per month per household, of which 50 cents would be revenues to the operator of the service. One configuration would be an extended prime-time news service—from perhaps 6 p.m. to midnight—which would contain news, documentary and public affairs programming at the national and regional level. Its costs would be higher than that of a conventional network news operation—perhaps $75 million a year. It would therefore require approximately 12 million households, or half the expected minimum penetration of cable television at the end of the decade. The second configuration might be a twenty-four-hour news channel, similar to existing twenty-four-hour news radio stations, over which there would be considerable repetition of national and international news, time allotments for local and regional news, and a budget of special features. Again setting costs at $75 million a year, a penetration of half the minimum cable households would be necessary. In both cases, however, the degree of penetration could be reduced if commercial announcements were added to the mix. Revenues for such advertising, set at $2.50 per household in conformity with present network expectations (and the figure is indeed substantially higher, since revenues from local advertisements at each end of the news program are not taken into account), would reduce the break-even point below 10 million homes.

In neither case, of course, is it to be assumed that a channel dedicated to news, at a fee, would have a constant audience of 10 million homes, or even any sizable part of such an audience. Like any special channel, it would be available when it was desired. A news-hungry family might make use of a news channel each evening; another family might be content to view the channel upon occasion, or when a particularly interesting documentary was being presented. The ratings game in cases of this sort, unlike the ratings game to which television usually appeals, deals with cumulative audience over a period of a month, rather than audience per program minute.

A Superior News Service

Each of these distinct configurations would provide, in its own way, a service far superior to the present network service; one because it would be richer in documentary programming, the other because it would be permanently available. The demand for news, moreover, is sufficient to give some reason for a belief that competing cable news services would come into being, and that during the decade, at least, the viewer would be able to choose from among two or three network news services and perhaps several cable news services. For a modest price to the viewer, in short, cable promises a greater diversity in international, national and regional news and news-related programming: a wider variety of topics considered worth covering; a wider range of views concerning what is being transmitted.

We admit that this statement of the promise of cable television in the area of hard news and documentaries may be over-optimistic. Yet the penetration we hypothesize, and the cost to the subscriber, are in each case far smaller than comparable figures for the daily press. The numbers we have put forth here we concede are large, but they do not appear to us to be outrageous.

Local news does not respond so readily to the promise of cable. The major costs in producing local news at anything near the level of technical and editorial proficiency with which the networks have made the viewer familiar are not substantially lower than ordinary network costs, and must be recovered on a far smaller economic base. The problems that afflict the local broadcasting television station, if it wishes to produce an acceptable local news program and local documentaries, are to be found in almost equal measure for the local aggregate of cable installations.

The most that can confidently be asserted is that the cable entrepreneur, whether operating on channels he himself controls or, more probably, by means of leased pay television channels, will enjoy a substantially larger cash-flow from news operations than the conventional television station enjoys, and may be able to provide a somewhat more acceptable local news service. The question is, as it should be, fundamentally one of how much residents of a large metropolitan area are willing to pay for the provision of a metropolitan news service. The answer to that question will come as individuals and groups of individuals assert themselves to discover it. What cable television provides, in this instance, is nothing more than the channel capacity to make room for the entrepreneur.

Although the matter is somewhat beyond the terms of reference of this Report, it might be mentioned that the development of Public Television, in the face of the growth of cable television, might well lead it in the direction of local news. The various special audiences that Public Television now serves with entertainment programming and with national news-related programming will in all likelihood fall to pay television. In most major areas, a Public Television station, by virtue of its broadcasting license, will be retransmitted on the cable. Its role today is to broadcast programs which are economically impossible or unlikely over commercial television; if its role in an era of cable television is interpreted in the same fashion, local news and news-related programs preeminently meet that criterion.

There is a particular flexibility on cable television with respect to what might be called "raw news" or "unedited news." The ordinary news service is a process by which professional newsmen make editorial judgments on the news that passes before them, on behalf of those who wish access to the news, selecting from all the events of a given time period those which it chooses to present and imposing structure on each story and on the presentation of the news as a whole. But there is also possible an unstructured presentation of news, in which it is assumed, for example, that the proceedings of the City Council are of importance, and those proceedings are transmitted in full; it remains up to the viewer to impose structure on the whole.

This chapter is directed to a consideration of structured news and, subsequently, structured opinion. But the power of the cable to transmit unstructured local news is limited only by the capacity of the installation and the rules under which the installation is operated. That power will be treated in subsequent chapters, where consideration is given to governmental and quasi-governmental uses of cable television, and to the provision of channels to which the general public and public institutions have access.

The Expression of Opinion

The expression of opinion is inseparable from the presentation of news or news-related programming. The editor, in determining which news story to present and which to cast aside and in imposing structure upon those he selects, exercises opinion; the correspondent does as much when he chooses from among those he might interview, and decides what portions of the interview he will register. The professional takes every precaution he can to diminish the effect of opinion upon what he reproduces, but in the end he must fall back on his professional judgment, which is to say his professional opinion. And in any case, what he does choose to reproduce will be, in many instances, the expression of someone else's opinion, because in the editor's judgment that opinion is news. As for documentary and public affairs programs, the expressions of opinion are likely to be their very essence.

The First Amendment to the Constitution, and the legal decisions that have flowed from it, establish full freedom of expression for the press. The government cannot interfere with the expression of opinion except in extreme cases involving such elements as incitements to riot or obscenity; even in extreme cases its powers are severely limited. As a general proposition, the owner of a newspaper can assert in its pages what he wishes to assert, even though it may be demonstrably untrue, or demonstrably unfair, or demonstrably unwise.

The classic argument for freedom of expression comes from John Milton: "And though the winds of doctrine were let loose to play upon the earth, so Truth be in the field, we do injuriously by licensing and prohibiting to misdoubt her strength. Let her and Falsehood grapple; who ever knew Truth put to the worse, in a free and open encounter."

The problems for television, and for radio before it, are implied in the last few words of Milton's statement. Because the printing press is in principle copious, and because accordingly one man's newspaper, or book, or pamphlet can be confronted by another man's, there is, again in principle, the possibility of a "free and open encounter." But it can be argued, and indeed has been accepted as a principle, that where the exigencies of electro-magnetic spectrum space limit the number of radio or television stations in any locality and limit the number of such stations in the nation as a whole—where, indeed, they lead to monopoly or oligopoly situations—"free and open encounter" on television ceases to exist in any real sense of the words, and "licensing and prohibiting" become appropriate.

Cable television, by freeing television from the limitations of radiated electro-magnetic waves, creates for television as a whole a situation more nearly analagous to that of the press. As we have speculated above, the copiousness of cable television makes it possible to conceive of far broader access to its channels by competing entrepreneurs and hence opens up the possibilities of a far broader expression of opinion. The existence of public access channels, and the recommendations this Commission will make concerning their general availability and the principles governing their use, will make possible the expression of an extraordinary range of opinion, in practice as well as in principle. Yet the question remains whether these capabilities that cable television provides will in fact result in "free and open encounter" by means of television. And the power of television is such that it can certainly be argued that television must be treated, in this respect, as an entity: a pamphlet is not entirely a useful response to a television broadcast or cablecast.

The Question of Fairness

Complications arise when the audience for the expression of opinion is considered. It is clear that the reach of an opinion expressed in the course of a popular entertainment program is far greater than the reach of an opposing opinion expressed over a public access channel: the first might, in the New York area, reach the attention of many millions of viewers, the second no more than a few hundreds. This is no more than is true of the printed word. But the range of cost between the elaborate program and the public access statement is immensely greater than the same range in the printed medium, and even the lower limits of that program cost put it out of the reach of all but a few principals. Money, in short, can buy audience on television in a manner that can bar all but the extremely rich or the extremely powerful from the effective expression of opinion.

The FCC has sought to handle this inequity, on radiated television, by the "fairness doctrine." The doctrine provides that where real differences of public opinion are manifest, the expression of one opinion on a broadcasting station must be balanced in some reasonably equitable fashion by the expression of other relevant opinions on the same station. A station that broadcasts commercials intended to stimulate the sale of cigarettes must make available time for those who oppose cigarette smoking for health reasons. A station that broadcasts its own editorials on highway construction, pro or con, must make available its editorial time-slot for those with other views on highway construction, con or pro.

The fairness doctrine ameliorates some of the inequity without eliminating it. It remains true that a cigarette commercial interposed in the telecast of "Bonanza" is scarcely balanced by an anti-smoking commercial in the intervals of a religious broadcast. But at least it is on the same station and in some degree has its chance to reach the same audience. There are no perfect solutions, and the fairness doctrine is perhaps the best that can be done.

The doctrine, however, has at least one disturbing side effect. It tends to make the station, and through it the networks, reluctant in many cases to express any opinion at all. There is always the risk that in presenting a documentary, or even a hard news program, the network or station will lay itself open to charges of unfairness, and that it will accordingly find itself obliged to allot time—its only salable commodity—to a compensating program. The result may be a general blandness, and the deprivation of the viewing public of access to any kind of opinion on a matter that may be of considerable importance.

The Commission did not seek to make recommendations regarding the applicability of the fairness doctrine to radiated television. So far as cable television is concerned, the Commission concluded that no such doctrine should be applied to the operations of public access channels, where it will recommend that accessability be ensured and where accordingly the conditions of "free and open encounter" will be fully met. Channels which merely rebroadcast radiated signals will obviously continue to be covered by the FCC fairness doctrine applicable to those signals.

On channels devoted to cable-originated programming, whether free or subscriber-supported, the majority of the Commission has chosen to await the lessons of experience. A system broken up into many intermediate-sized audiences might require no fairness doctrine at all, upon the assumption that diversity of choice or program would bring about diversity of flow of opinion, and that in any case the accessibility of public channels reduces the need for regulatory intervention. A system in which a small number of channels commanded the great mass of the audience might create the fears of monopoly of opinion to which the FCC has reacted with respect to conventional television.

Questions of ownership and control of cable installations are also relevant to questions of fairness. If the mass of individual cable installations, throughout the country, were to be owned or controlled by a few large corporate enterprises, as networks are today controlled, the spectre of monopoly of opinion would arise in quite a different form. Still another form of monopoly would arise if, within a given geographic area, many communications media, including television and cable television, fell into the same hands. The Commission believes, however, that its recommendations on ownership and control over cable television preclude the likelihood of such outcomes.

CHAPTER NINE

PUBLIC SERVICES ON THE CABLE

Preceding chapters have dealt with services that are, in effect, extensions of conventional television; they constitute an attempt to examine how conventional programming might alter in a television of abundance. There was an assumption that a market could be provided in which the expanded service could be financed on its own economic base, drawing, directly and indirectly, revenues from audiences in many cases substantially smaller than the total audience to which conventional television generally addresses itself.

This chapter, in contrast, deals with a quite different family of services. They are the services which directly address the public interest. Primarily in the area of information, they deal with the *need* to communicate rather than the desire to profit from the by-products of communication.

They are not necessarily communications on the large scale. Concerned with such matters as health and welfare, the interaction of local government with its constituency, social and economic needs of the community, and the relationship among the many sub-groups that make up a large city, their range may

97

be primarily local and at its broadest extension not likely to cover more than a single metropolitan area of a single state.

It is, of course, the abundance of channel space characteristic of cable television that makes it possible to conceive of providing such services. Even at twenty channels, and far more so at forty, the demands of the total audience and of the large sub-audiences are not likely to occupy the entire cable capacity. The remaining channels, under any system of regulation, are available for the best use that can be made of them.

But there is also something more than abundance that makes cable television so well adapted to the provision of such services. The configuration of a cable system, as an earlier chapter pointed out, is that of a tree: the signal travels up the trunk and out through the branches. And like a tree, a cable system has a maximum size, beyond which the laws of physics and the laws of economics make further growth impossible or impractical. The area served by a single system is measured in a small number of square miles within which, in the most densely populated sections of a large city, there may be found perhaps 100,000 residences. As one moves away from the inner city, the optimum size of the system becomes smaller with reduced concentration of residential units.

Systems can of course be linked, head-end to head-end, by a variety of means to make of all the individual systems a single huge system, or in any other configuration one might desire, just as railway cars are linked to trains. But just as the fundamental unit on the railway is the car and not the train, the fundamental unit in cable television is the individual system. Whatever other capability it may possess it is able, as conventional television is not, to serve its own community and that community alone.

Trial and Error

This chapter deals with the horizons that are opened by the enormous capacity of cable television to serve the community, or sets of related communities. For the Commission, it has been a significant portion of its efforts and a portion that is clearly worth pursuing, however dimly its potential may present itself. Conventional television has seized a communications instrument of unparalleled power and applied it to serve the nation's requirements for entertainment and general information; the physical characteristics of broadcast television and perhaps our own national characteristics have made that outcome inevitable. (Others are now seizing that same instrument, with its capacity to carry data, and turning it to the needs of commerce and industry, to serve one segment of the public's special requirements.) But cable television has the capacity to provide more. It can affect much more basically how people live, their health, education, their jobs and their community cohesion. Unfortunately that kind of impact still defies specification: we can only guess ahead and plan. The experiences with this kind of use in so new an enterprise are necessarily minimal.

In dealing with the uses of cable television for the mass audiences that the new technology is able to serve, the Commission was able to rely to some extent upon long years of experience with broadcast television, itself addressed primarily to mass audiences. In considering the uses of cable television in the public interest, the experiences of the past are less instructive.

There have been experiments in the use of cable television for the provision of information concerning such matters as health services, employment, consumer education, formal instruction and community development. On the whole, however, there is little to be learned from these isolated and sporadic experiences. Cable television as a whole has been a small enterprise. Most systems are extremely small systems, in small communities; they do not provide an economic base upon which to provide much more than the retransmission of broadcast signals. The large cities, where public interest uses are of the greatest significance and can make the greatest difference, are as yet virtually unserved by cable television.

Under such circumstances, an analytic study of the uses of

cable television is beyond the power of the Commission, its staff or its consultants. The best that can be done is to attempt to match current needs with foreseeable technology, with all the uncertainties that such a synthetic approach imposes. In the end, these uses of cable television will be worked out by a method of trial and error; that method is not yet available to the Commission and indeed, except in special circumstances, not yet available at all.

In its deliberations, the Commission proceeded by establishing with the assistance of technical consultants what it believed to be a reasonable technological basis. That basis was then submitted to experts in the various public interest fields that appeared to be in some degree highly dependent upon communications, and on that basis they were asked to specify, so far as it lay within their power, the uses within their own fields to which such a cable system might profitably be applied.

Their reports are in hand, and can be made available to interested persons or agencies.[1] Included are extensive speculations on the use of cable television in the provision of health services, education, employment services, community development, consumer education, services to the labor market, services to the welfare recipient, and related subjects.

A Possible Menu in Bedford-Stuyvesant

The Commission has also profited from an engineering and management study directed toward establishing the feasibility of a community-controlled cable television installation in the Bedford-Stuyvesant area of Brooklyn, N.Y. The study was prepared by Malarkey, Taylor and Associates on behalf of the Bedford-Stuyvesant Restoration Corporation; a section is devoted to brief descriptions of individual programs or program series

[1] Abstracts appear in Appendix E.

that might be contemplated for subscribers. Some of these are entertainment programs, but included are also brief scenarios for the following:

—Job-A-Rama: Job opportunities, instruction in job interview techniques and preparation of application, other employment services.
—Children's Playhouse: A light educational background for pre-school children.
—Area Center Parade: Explanation, documentation and advertisement of community center activities, transmission of special programs instituted by the various community centers.
—Street Scene: A roving-reporter presentation of "what's going on in the community," as a means of building community identity.
—The Consumer: Bargain hunting, shopping techniques, money-saving hints.
—Kings County Hospital: To create wider attention of endemic health problems and to assist in methods for their eradication.
—Brooklyn College Journal: To provide training, on the cable, in all phases of broadcasting, from administration to production.
—Restore Digest: Community self-help programs prepared by the Restoration Corporation.
—The Drug Scene: Documentaries pointing out the dangers of addiction and the roads to rehabilitation.
—The Black Man: A programming effort highlighting the Black cultural heritage.
—Gospel Hour: Incorporating religion and music into the same production format; reinforcing the activities of the 300 store-front chapels in the community.
—English Lessons: For those conversant only in Spanish.
—Pratt Institute Hour: Devoted to artistic instruction, based

at the local Pratt Institute, which specializes in such
instruction.

Beyond these specific program notions, the study proposes
in more general terms channel usage by the Fire Department,
the Police Department, the Recreation Department and so forth,
and the installation of a small computer which would permit
use of the system for game-playing.

The Commission list is nothing more than a list; the Bedford-
Stuyvesant brief scenarios are merely scenarios. They become
presentations only when time and attention are devoted to their
preparation, when they are presented on the face of a television
set, and when their acceptability has been demonstrated. That
they, and the whole cause, are worth time and attention appear
to the Commission to be beyond question.

"Health Services" appears on the Commission's list; "Kings
County Hospital" in the Bedford-Stuyvesant study. Some further
indication of what they might involve, on both the national and
the community scale, might be instructive.

Health Services

The role of communications in the provision of health services
falls into two natural divisions. One of these concerns communications at the community level, between those who provide
health services and those who receive them. The other concerns
communications on a broad scale within the profession of medicine, among the professionals and para-professionals who provide the services.

It is being increasingly recognized that serving the informational needs of the public is an important aspect of health-care
delivery. To the extent that individuals are well-informed concerning organized medical service, and to the extent that they
are prepared to cope on their own with health-related problems

—nutrition, child-care, chronic disability, family planning, safety within the environment—the entire health care system is strengthened. It has been, in the past, the lack of any effective mode of conveying such information into the home that has been the principal constraint on such public education. Cable television can serve to provide that effective mode.

Health educators have utilized broadcast television whenever the good will of individual station operators has permitted them to do so. Most stations cooperate when an emergency threatens their community, as for example in the aftermath of a hurricane or flood, or when some such health campaign as a drive for Rubella vaccination is in progress. But in each such instance the health educator must accept sporadic and off-hour time-slots for his spot messages. His flexibility is highly constrained, and his audience limited; the amount of information he can convey is far from consonant with his need to convey what may be a somewhat sophisticated message.

One can readily identify four areas in which cable programming, on a steady basis, could make an important contribution. The first of these is the area of *health care assistance.* It would incorporate programming directing individuals to care for themselves in illness. Some of this would be of broad general interest, such as a series on the medical and non-medical uses of drugs, including poison control. Other programming might be directed toward the elderly, the chronically ill, the pregnant woman or the young mother. Question and answer programs, using the well tested method of the telephone call-in, could stimulate a miscellany of information on common household diseases, accidents and general health programs.

Preventive programming might present basic practices in preventive health measures. A series on the preparation of low-budget, high nutrition meals would be of tremendous value to the ghetto housemaker. In the area of preventive dental health, a similar series of programs has already been proposed in the Watts Community of Los Angeles. Hygiene, family planning, sanitation, mental health all lend themselves to such an approach.

Medical practice orientation would provide a continuing guide to the intricacies of medical organization. Lack of knowledge about the hospital system, and the consequent fear that may be aroused, are often deterrents to the procurement of services.

Finally, there is a broad range of *community health* information which might uniquely be delivered by a cable system. Information about Medicare and Medicaid is of prime significance. An effort might be made to provide current information about local health services on a continuous basis, stating times and places at which various health services are being provided, or mobilizing a community for health-related campaigns. In particular, such a system might well be linked to local medical vans or storefront installations, to bring the hospital into the community either on a recurrent basis or for *ad hoc* purposes.

It might be said that by means of the last two of these services, the community hospital would possess for the first time a promotional medium that would enable it to sell its wares within the community. Without such a medium, the underprivileged in large cities remain simply unaware of what is perhaps one of the most important resources within its boundaries; the hospital is merely a point of last resource in emergencies. Medical assistance provided in such a manner is performing only the least of its tasks.

Such programming is never likely to attract large audiences, except perhaps in periods of epidemic or catastrophe. But the value of a health service of this sort, and indeed of any public interest cable television, is not measured as broadcast television is measured, in size of audience. The proper calculus sets the value of the services to those who benefit from them against the cost of providing those services. If preventive health education, in a community of a few hundred thousand persons, maintains the health of a few thousands of the population—that is, if the effective reach of the program is calculated at 1 per-

cent — the cash value of the programming may be in excess of
$1 million. Such a return would satisfy the most exigent ac-
countant.

Continuing Education for Doctors

The second area in which cable television might contribute
to the provision of health services concerns the physician rather
than the patient. The medical practitioner is the product of long
years of professional training. At the conclusion of that period,
he typically establishes himself in private practice, and sets
about applying to the needs of his patients the knowledge and
the skills he has accumulated.

But the practice of medicine is intimately linked to medical
research. It progresses with the progress of the research itself.
As the period lengthens between the doctor's formal education
and his current activities, the process of diagnosis improves,
the procedures of choice are altered, the very definition of medical need takes on new meaning. From these developments, the
practicing physician is largely isolated. Without distinct effort
on his part, he is condemned to go on doing what he was taught
to do at the beginning of his career.

In the past, major efforts have been made to provide appropriate services to the practitioner through the use of film. The
largest such enterprise has been conducted by the National
Medical Audiovisual Center of the United States Public Health
Service. NMAV has provided, on a loan basis, a wide variety
of medical films and other audio visual materials to health institutions throughout the country. In the private sector, the
Roche Laboratories have fulfilled a similar responsibility, distributing a bi-weekly journal of films to some 600 hospitals
and medical schools and making old films available for replay.
Some of the larger medical associations distribute smaller
quantities of such programming.

But while these activities appear to reflect the demand for

such activities, they do not solve the fundamental problem of distribution. The practitioner must still get to the place where the material is shown on a schedule which is not of his devising, and for a reward which may prove something less than satisfactory after the first few minutes of the film have been shown. Under the circumstances, it is hardly surprising that the vast majority of doctors are never reached by such services; those that are reached are not particularly well served.

Transmissions by Public and commercial television have been tried. But even where they have been reasonably well received, the principal achievement has been to demonstrate the incompatibility of broadcast television with continuing medical education. Limited access and off-hour time-slots simply cannot attract large numbers of professional viewers with any degree of regularity, nor is it economic to supply such an audience on the scale required for high-quality service.

The major problems, in all such experiments, have been those of access and scheduling. Those are precisely the problems that the cable television system can most directly attack. At high penetration, the cable system will reach the doctor wherever he may find himself with the time and the desire to turn to self-improvement; the set in his office during a quiet interval will do as well as the set at home after dinner. He can, at such moments, switch to the medical channel (where such a channel exists), determine whether the information being transmitted is of interest to him, and watch it or not as he pleases. Alternatively, he can check the schedule of medical transmissions and arrange his home and office day so that he may watch them. Further flexibility is provided by steady repetition of programs, providing him with a choice of time periods at which to tune in.

The problem that remains to be solved is that of providing first-class programming. It is not surprising, for the programming that now exists serves only a few thousand physicians in total; there is a limit to what can reasonably be spent to reach so small a target. The future of continuing medical education, like so

much of the future of cable television, depends upon the growth
of the system.

The remarks made here concerning physicians apply with
equal or greater validity to the para-professionals with whom
the physician deals. Nurses, ward attendants, and technicians
require continuing education and re-education quite as much
as the doctor, and have at present even fewer facilities for the
satisfaction of their needs. The same system that serves the
professional can serve these people as well; it is hardly necessary to elaborate the manner in which that could be accomplished.

Cable and Formal Education

The potential of cable television in the service of formal education—that is, as part of the school and higher educational
system from kindergarten onwards—has been universally acclaimed. In the early days of radio and of broadcast television,
the potential of those media was equally acclaimed. Yet neither
medium has made any significant difference to formal education,
and the difference would be negligible if both were suddenly
to disappear. It may be that that sweeping statement will require modification if the Children's Television Workshop provides a successor to "Sesame Street" that will be effective for
first-graders; it is attempting to do just that, and its previous
success with pre-school-age children gives considerable reason
for confidence in the outcome. But the record to date is dismal,
and even with the addition of a first-grade "Sesame Street" the
bulk of the system will have been largely unaffected.

It appears at first sight absurd that formal education, itself
primarily a communications process, should be so indifferent
to the provision of powerful new tools of communication. But it
is conceivable that it is precisely the power of the new tools,
and in particular the power of television, that makes them

appear incompatible with the existing educational system. For while television can be added to the existing health service, for example, and at least initially leave that system essentially unchanged, its full impact on the educational system might be enormous.

Despite the record of the past the Commission is convinced that cable television has a role in education, and perhaps a role of surpassing importance. Its advantages over broadcast television lie primarily in the abundance of channels, which makes it possible for a community to enjoy (if it proves worthwhile) a channel or more for each grade, if grades survive. The digital return system may be of great significance, and further elaboration of return signals with the passage of time of even greater significance still.

But the Commission believes also that a study of the significance of cable television (and closed circuit television) in education must be undertaken as part of the study of the educational process as a whole. The power of "Sesame Street" derives from the fact that it was preceded by a serious study of the educational needs and the educational processes in pre-school children. The problem is not sensibly to be considered as a question of what cable television can add to the formal educational process, but as the broader question of how the entire process changes when a powerful new tool of communications is added. It may well be that in the light of the new tool, every major aspect of the system requires significant change; perhaps if one adds television, a whole new armory of textbooks must be provided, a new kind of schedule devised divided between schoolroom and home, new subjects added to the curriculum and old ones discarded. The Commission is not implying that it believes any or all of those statements to be true; only that the study of the educational system must extend to those limits and beyond.

Because the Commission was itself not prepared to embark upon such a study, in itself far larger than the study it undertook, the Report makes no direct recommendations concerning cable

television and education beyond urging that a major study of
the sort suggested above be undertaken, and that provision be
made for enough channel space for education, even at the present time, to allow for experimentation. These comments should
be viewed in the light of our conviction that the matter is one
of pressing importance, and is not treated summarily in this
Report out of diffidence or indifference.

Closed Circuit Possibilities

The power of cable television is its ability to reach into homes,
and to provide large numbers of people, at moderate cost per
person, with entertainment, information and services. The
system is most fully and most economically exploited when it
is used in that fashion.

While the system grows to maturity, however, and particularly in those franchise areas where the system operator has chosen
to install forty-channel capacity or more, whole channels are
likely to remain idle or grossly underused over a period of several years. Under such circumstances, it is economical and potentially useful to employ the redundant channels as closed circuit
systems; that is, systems in which communication is provided
only at a small number of locations. In a large city, the redun-
dancy might be used, for example, simply to link up police
headquarters to a few dozen precinct stations, or a central studio
to several dozen day-care centers, or perhaps all schools to
central educational facilities.

In the long term, as demand grows for the cable system
channels, such uses might be more economically served by installation of closed circuit systems, serving the selected points
and no others, or perhaps better still by point-to-point micro-
wave services. In the interim, however, the cable system at no
cost to itself or to its normal services might be of major assistance to the various institutions of government, and to some

non-governmental institutions, to which the availability of a sophisticated communications medium at low cost might be of great value.

Uncertainties in the Delivery of Health Services

To deal in some detail with health services, dismiss almost summarily formal education, and make passing reference to closed circuit uses, is scarcely to cover the range of services that cable television might provide the community. But there is a limit to the number of topics that a Report of this kind can encompass, and a certain doubt that it is wise to attempt too much when the discussion itself must be frankly speculative.

The essential point is that almost wherever communications is a vital part of a process, the cable television system's new potential is relevant. If one believes that personal security is better maintained by possession of a burglar and fire alarm, that alarm can be provided economically and efficiently by cable television. In an area where jobs are looking for men, and men for jobs, cable television can assist the process that will bring them together. The social worker can find cable television useful in the management of the welfare system; it can be an important medium in the housing market; it can be used to allay rumors in volatile areas. As an instrument of consumer education, it may prove more successful than media now available. It can contribute to neighborhood cohesion by transmitting local sporting events and other entertainments. All of this, and more, lies within the capabilities of the system.

It is a far cry, however, from stating capabilities to realizing them, and there are several obstacles to be overcome if the public service uses of cable are ever to be exploited. First, the physical capabilities of which this Report speaks do not as yet exist. There is no highly penetrative twenty-four-channel or forty-channel, digital-response cable system, and whether one will ever come into being will depend upon the decisions of Congress and the regulatory agencies, upon the willingness of entrepreneurs to provide the necessary venture capital, and upon the general public's willingness to install the new instrument, at a price, in their homes. The political decision will be particularly difficult, since it is likely that cable television will unsettle existing arrangements.

Second, there is no assurance that messages which will be both informative and persuasive can be prepared. Few agencies have tried on any ambitious scale to use cable (or television, for that matter) and the attempts, if they are made, will be costly and often frustrating. Whether funds will be available in an era of tightening government budgets and rising government costs is again beyond our capacity to predict.

There is no guarantee that the extra channels cable can provide for public services will ever command audiences sufficient to make the enterprise worth the investment of scarce resources. It is true that audiences may be sufficient even when miniscule by the standards of entertainment television. It is also true that in many cases the character of an audience will be more significant than its size. Nonetheless, it is indisputable that a health department, for example, will not know what to expect until it has tried, and that trying (and trying again) will be no simple task.

Finally, there are special risks when the government controls television channels. In the past, because channels have been scarce, the risks have been so great that government uses have been exceedingly limited. For the provision of government services is never a neutral issue. Health and welfare, for example, are important public issues; and information about them can easily become advertisements for incumbent administrators.

Yet, in spite of all this, the Commission is convinced that the venture is well worthwhile. The conviction rests upon an awareness that cable, in many important respects, can become a powerful medium, and upon a faith that society will inevitably find the means to turn such a medium to the service of its own best purposes. The Commission convened with the belief that cable has much to offer toward the solution of certain social problems, and in particular certain urban problems. Nothing it has learned, during its months of deliberation, has shaken that belief.

Preconditions

Clearly, there are preconditions which must be met if the uses of cable installations in the public service, by governmental and quasi-governmental bodies, are to be exploited in a fashion that will serve the real needs of their communities. There must be, in the first place, channels set aside for these uses, in a fashion that will maintain them immune from incursion by commercial interests. The Commission will make specific recommendations concerning the allocation of cable channels; they are better deferred, however, until the full configuration of the cable system has been explored.

There is also the necessity that cable service be available to those portions of the city which particularly require special services. This will not occur automatically: cable installations are most likely to be made first where the promise of economic return is greatest. To the Commission, these appear to be primarily issues of franchising and other regulatory policy; they, too, will be considered in their appropriate place within the Report.

There must, finally, be government planning and government experimentation on a large scale. A coordinated federal initiative, including special funds for experimentation and use of cable, is essential if the new technology is to be turned to the uses of the cities. Government has been reluctant to use scarce television for its own purposes; in the television of scarcity, the notion of a government channel was anathema, for more than doctrinaire reasons, if the alternative was entertainment. In a television of abundance, it should be assumed from the outset that government uses will be heavy, and indeed essential for the optimum growth and the optimum use of the system.

The federal government already possesses the instruments by which it may make itself felt in the development of cable television. An Office of Telecommunications Policy has been formed and charged with responsibilities which include oversight of the size and direction of the growth of cable installations. A Task Force on cable has been formed which includes, among others, the Secretaries of Health, Education and Welfare, of Housing and Urban Development and of Commerce; to it might appropriately be added a representative of the Office of Economic Opportunity in its new role as research arm of the government to explore promising areas for social experimentation. Agencies themselves, and in particular HEW, HUD and Commerce, have begun to alert themselves to the potentialities of cable television, and can be expected to make funds available for the encouragement of new enterprises; that interest should be stimulated.

Redistribution of Population

This chapter has dealt, in general, with services to the city, and to the communities that constitute the city. There will be those who think our view too pedestrian, for there is a larger view of cable television that can be embraced, in which the tactical steps this Report recommends give way to larger strategic consideration. The most persuasive spokesman for that view is Dr. Peter C. Goldmark, and a strong statement of that view has been made by a committee of the National Academy of Engineering on which Dr. Goldmark served.

That committee sees cable communications in the broad sense (as distinguished from cable television as it is considered

in this report) as a means of accommodating the growing population of the United States to evolving patterns of employment,
and without contributing to the cancerous growth of existing
centers of population. They envisage enormous networks of
cable communications developed in a fashion that will fill many
of the communications needs that must now be cared for by the
physical movement, and the face-to-face confrontation of those
who command and must use information. In such a fashion,
population can, they believe, be more sensibly distributed, and
some of the amenities of human life restored.

The Commission was impressed when Dr. Goldmark appeared before it, and impressed again when it received the
report of the Academy. We certainly believe that the view he
expresses, and which indeed others have expressed, should be
taken seriously, and their implications carefully studied.

At the same time, however, we believe that the time-span
they indicate is greater, by some large measure, than the time-
span upon which we have chosen to work. The sheer size of the
capital investment that would be required assures that no such
major broadband network is likely to come into being within
the decade, or even within the century. Meanwhile the steps
we recommend are consistent with all that these broader programs envisage, and can be taken without prejudice to more
ambitious undertakings. Whether cable communications can
transform urban life, and with it life in the United States in
general, remains for this Commission an important but an open
question, with which we have not sought to deal. That cable
television can contribute to urban life we believe to be highly
probable, and we urge that every attempt be made to find out.

CHAPTER TEN

POLITICS ON THE CABLE

An earlier chapter of this Report deals with the expression of opinion on cable television. It can be taken for granted that much of that opinion will be political opinion; it will imply some kind of governmental action and by indirection the election to positions of power of those who share that political opinion. But such expressions can be differentiated from forthright political messages, which are explicit calls to action, or explicit support of a specified candidate, or invitations to adhere to a given political party or a given political attitude. The political process, in other words, can be seen as an activity in its own right — a part of the general expression of opinion, to be sure, but possessing characteristics that most of the time demarcate it with some precision.

Over the past decade or more, broadcast television has become the medium of choice for the delivery of political messages

NOTE: This chapter draws heavily on a paper entitled *Politics in a Wired Nation* prepared for the Commission by Ithiel de Sola Pool and Herbert Alexander. The full text of that paper is available on request from the Alfred P. Sloan Foundation.

to political constituencies. In 1968, political costs at all levels, in primary and general elections, were in the neighborhood of $300 million. During that year, political broadcast expenditures, including production costs and agency fees, were approximately $70 million. To this figure must be added at least two others: the cost of "tune-in" advertising in the daily press and other promotion expenditures, and the cost of staging political events of which the principal purpose was not the event itself but the possibility that it would produce for the candidate exposure of a national or metropolitian television newscast. Thus at least one-quarter of all political spending is directly related to broadcast television and radio, with television taking by far the largest share. No other medium begins to approach this figure; newspaper advertising, for example, came to something in the order of $20 million.

The attractiveness of broadcast television is at once obvious. It has demonstrated itself to be the most persuasive advertising medium the world has ever seen. It can provide immense audiences, attracted to the set by the entertainment with which the advertising message is surrounded. The advertising message itself is not easily ignored, particularly if it comes in the form of a thirty-second or one-minute "spot announcement." Ordinary inertia on the part of the viewer is an assurance he will be exposed to the message; there is no such guarantee in a newspaper advertisement, which must attract the reader on its own merits. There are even those who will adduce figures to show that the inertia of the ordinary viewer is so great that he will tolerate a thirty-minute political program if it comes between two attractive entertainment programs. At campaign time, such thirty-minute segments, and more, can be purchased for a price.

There are, indeed, those who claim that the enormous persuasive power of broadcast television constitutes a real threat to the political process. That enormous power is associated with enormous cost. The candidate with bulging coffers, it is argued, possesses an unassailable advantage over the candidate who finds television beyond the reach of his personal fortune and his fund-raising capability. There may be dangers that the primary and the election can be bought. There was, in 1968, some evidence that in certain situations such an outcome indeed occurred.

One limitation on this process is of importance. It favors the candidate for national or state-wide office over the candidate for local office. The broadcast television system is not designed to reach the ward or the precinct, or even the municipality; the signal emitted covers an area measured in thousands of square miles, and the cost of buying time on the system is scaled to that kind of audience. A candidate for municipal office in a Connecticut suburb would be obliged, if he wished to campaign by television, to employ a system with a potential audience of some 20 million viewers; to reach the few thousand viewers he seeks to reach, he must pay for reaching all 20 million. At the municipal level, this can rarely be attractive. Even at the state level, it may be inordinately expensive, as when a candidate for Governor of New Jersey is obliged to buy time on New York City stations to reach a large part of his constituency.

Matching Audience and Constituency

Cable television, by its physical configuration, eliminates some of the disabilities of broadcast television. With channel space to spare, time even on a state-wide basis need not be expensive; the purchaser is not necessarily in competition with the mass advertiser. But what is more important, the cable system is so organized that a political message may be directed to a potential audience more or less identical with the constituency of the candidate: to the ward or the suburb or the state. Each person the message can reach is a person who will be casting a ballot of significance to the candidate who buys time and pays the cost of producing the message. The problem of spill-over becomes insignificant; the costs associated with spill-over vanish.

All this comes into existence without diminishing the power of the present system. The cable television installation will continue to carry channels of mass entertainment, and most of the audience, at any time, will be tuned to one or another of those channels. The candidate who wishes to pay the going price for a spot announcement or a time-slot on those channels can continue to do so, and reap the demonstrated benefits of his choice. What cable television provides is not a substitute for the current mode of political promotion, but a supplement. It can perform all that conventional television performs—and something more.

The manner in which this supplementary service might develop will be, once more, the consequence of a period of trial-and-error. This Report can do no more than suggest some of the possibilities.

The most straightforward approach would be the institution, on a cable system or a set of such systems, of a political channel operating on a regular basis and both producing its own political programs and allotting time to the local, regional and national political parties. Such a channel might be organized and managed by a non-partisan local institution created for the purpose, along the lines of the League of Women Voters. Its viewing audience would rise and fall with the political temperature of the community it served: high during the general election, higher still if a local recall election was exercising the community, attractive most of the time only to the politically sensitive.

Partisan Channels

Beyond this simple system it is possible to conceive of the institution of explicitly partisan channels. Since the early days of the Republic, when all newspapers were party- or factionallyrun and oriented, the United States has never had an explicit party press. Costs of running such a channel, even on cable television, appear at first glance to be high: if it is estimated that delivery costs via cable come to approximately one cent a day, and a party committee wishes to reach 1 million households during the two-month period before a general election, delivery costs alone would come to $600,000, or approximately 20 cents a voter. The sum is not large, but large enough to make it necessary to reduce most other campaign expenditures. On the other hand, such a channel would be an extremely valuable fund raising instrument, and might well pay its own way.

The question of fund-raising, in fact, arises in its own right even in the absence of party channels. The merchandising capability of an addressed cable system with its capacity for a digital return signal makes it possible to address appeals directly into the home, and either to receive pledges in the system itself or to mount follow-up campaigns that might be extremely efficient. The same announcer who sells kitchen appliances on broadcast television might be available to say "We need funds and need them badly: push the button once for every dollar you want to contribute." Impulse giving may prove to be as profitable to the party as impulse buying to the merchant. "Political parties," to quote our consultants, "may find themselves more dependent on charismatic pitchmen than on a few rich men."

Still another possibility for the campaigner lies in the possibility, suggested in the previous chapter, that the mass audience portion of the cable system may come to provide an all-news channel comparable to those that now exist on radio. Exposure on such a channel, particularly for the local politician, will provide opportunities not now available on broadcast television. It can be a powerful instrument in making himself known to a wide public in a fashion that no other news medium can provide.

All of these are modes in which the campaigner and his party can direct their message and their appeals to the general public. The degree to which they reach the public will depend in large measure upon their skills in presenting themselves and their causes. In a certain sense, they will be competing with relatively scarce means against the mass entertainment that is available on the popular channels, and their audiences will not, at any given moment, be large ones although their cumulative audiences — which essentially is what the advertiser strives for — may be immense.

But there is another kind of audience which is far more dependable, and in some ways even more important to the campaigner. That is the audience, small in total, which is composed of the political enthusiast, the party faithful. It is those people who fund campaigns, promote the candidate of their choice, solicit face-to-face the support of the less dedicated, get out the vote on election day. In a political campaign, much of the effort of the candidate is directed to mobilizing those forces, and maintaining their initial enthusiasm over the long pull.

Closed circuit television, in which the candidate addresses several large gatherings in separated locations, is now in frequent use for that endeavor. Cable television is ideally designed for the purpose. It can reach the campaign worker in his home. There is no problem of attracting that audience; it is eager to participate, anxious to receive the word. Cable television cannot take the place of the candidate's handshake, and a few kind words about the family, but it can be an efficient and an economic supplement.

Exposing the Political Process

So far this Report has considered only the contributions cable television might make to the management of political campaigns. The use of the medium in exposing the political process itself, particularly on the local level, is at least as significant. It requires little explanation.

With channel space available, the problem of permitting the

community to participate vicariously in the political process itself
vanishes. The public hearing which is so fundamental a part of
local and state politics can become, in another sense, a public
hearing indeed, in which the general public can at least be
present, if only passively. City council proceedings can be
watched by those who are interested in seeing their government
in action. The school board can be forced to debate its several
positions in full view of those who will be affected by its decision.

Audiences for such activities will at all times be small except
perhaps times of municipal crisis. But the importance of such
an audience is not measured by its size. Those who listen tend
to be the activists within the community, who themselves take
a deep interest in the political process and its outcomes, and
who possess a deep personal commitment to the process. They
are the enthusiasts, whose principal purpose is to make their
own enthusiasms contagious. We have seen, in recent years, a
handful of environmentalists profoundly affect the national
view of the environment. In a small city or a suburb, a similar
handful who are interested in their schools, and who have via
cable television regular access to information about the management of their schools, can infect the entire community with
pride or with concern, and can bring about profound change.

The effects of all this on the entire practice of politics might
conceivably be far more far-reaching than the single items in this
catalogue suggest. As stated in the paper prepared for the Commission by Ithiel de Sola Pool and Herbert Alexander:

> If the local cable system serves to delimit neighborhoods,
> to give a sense of community to a section of a city or to a
> suburb now mainly dependent on the central city media,
> then politics could become more decentralized, with less
> attention to the national and state, and more to the local.

So far as conventional television is concerned, the FCC and
Congress have wrestled at intervals with what is known as the
"equal time rule." In principle, this provides that time allowed

any single political party or its candidates must be made equally available to all political parties and their candidates (except on hard-news programs); in practice, it has led stations to refuse air-time to any party or candidate, except on a commercial basis, for fear of invasion of air-time by every political party, however small, in existence within the state in question.

The Commission believes that if leased channels and public access channels are effectively available and employed the necessity for "equal time" will be removed, and that whatever its status with respect to conventional television, "equal time" need not be regulated with respect to cable television.

The existence of an informed and an engaged public is prerequisite to the healthy maintenance of a political democracy. Cable television, on the face of it, has much to contribute. Unlike conventional television, it does not necessarily call for the investment of huge sums, and it can be made effective down to the local level.

It cannot, of itself, create a politically aware citizenry, for no one can be forced to twist the dial to the channel carrying political information or political news. But cable television can serve, as perhaps no medium before it has been able, those who wish to be part of the political process, and skillfully used it might very well be able to augment their number. Like so much that has been put forward in this Report, its effectiveness will depend upon the skills of those who take it in hand. Politics, whatever opprobrium may sometimes be attached to the word, is important. The cable can literally bring that fact home, and in doing so help the entire political process function efficiently and effectively in the public interest.

CHAPTER ELEVEN

THE COMMUNITY VOICE

One large family of possibilities remains to be considered. These are the uses of cable television as a medium for the direct engagement of people with people; as an institution within which the separate voices of the community may be heard.

In principle, and to some degree in practice, the printing press has always been such a medium. Any person with a message to convey, and with the necessary modest funds, can have that message printed and can distribute it by his own means or pay to have it distributed; can, indeed, exact a fee from those to whom his message has appeal. Within the limits of obscenity, libel and the like, he may say what he pleases. It is nobody's business but his own if his message represents that the world will come to a sudden end in March, or that Bacon wrote the works of Shakespeare; he is under no obligation to be sensible, or rational, or popular. He may say what he thinks, and take his chances of finding an audience.

If the cable system is not as copious as print, it remains copious beyond anything that broadcast television might suggest. And

123

to some degree, the difference in capacity between the press and cable television is compensated for, and perhaps more, by the power of the television medium.

There are in every community issues and enthusiasts for those issues. Some of the issues, no doubt, are trivial or inconsequential. Many are irrational. But the test of the issue lies in its fate upon exposure, and the health of the community, in many respects, depends upon the ability of the enthusiasts to test their issues by exposing them. By making the exposure possible, in a potentially telling form, cable television serves the public interest.

The opportunity for those with issues and grievances to expose them progressively to larger and larger audiences is of particular significance in the inner cities, and it is there that television is not only the medium of choice, but (along with radio) the only pervasive medium that is available. The dilemma in such areas is exemplified in some of the consequences of the decision, in New York City, to decentralize control of the schools. Dissatisfaction with the operation of the school system is pervasive in New York City, but there has been no way for the community as a whole to examine the roots and the nature of that dissatisfaction, to reach a consensus on the manner in which it can be remedied, and to act on that consensus. In the absence of such a consensus one is likely to find that little emerges other than groups of warring factions, no one of them representative of the community as a whole simply because the community cannot be involved in their deliberations. School board elections attract only a few percent of the voters; the results of the elections can hardly be called representative, and the grievances remain not only unresolved but largely unstated.

A Need to Be Seen and Heard

But the case need not be made exclusively in terms of issues and grievances. There is a need, in every community, for the expression of common notions, for the expression of artistic and cultural endeavors; a need to serve the elderly, the deaf, the very young; a need for an audience that finds a resolution, in more affluent areas, in the creation of Little Theater Groups and similar associations. There is the need to express oneself in forms that can be carried across boundaries to similar communities elsewhere, and indeed to dissimilar communities which might profit from the expression of unfamiliar views. There is a pervasive need, in short, to be heard.

There is a special need that is not always considered in this context: the need for the entrepreneur to make himself visible. The local businessman within the large metropolitan area is in an almost impossible position: he has no method within his community to advertise his wares. The growth of small enterprise into large enterprise is blocked. Even the institutionalized undertaking suffers from the same deficiency: one member of this Commission, presiding over a sizable, non-profit mortgage fund in the Bedford-Stuyvesant area of New York, is unable to place mortgages even at rates substantially lower than those mortgage brokers within the area, simply because he has no way of making known the existence of that mortgage fund.

The results of this deficiency are manifold. The growth of economic activity indigenous to the area is precluded. The field is laid open on equal terms to the scrupulous and the unscrupulous, to the responsible and the irresponsible, with consequent increases in real costs to purchasers.

Two or more community channels, open to the members of the community for whatever their purposes, might go a long way toward relieving the pressures that arise where communication is in short supply. They will not solve the problems of the inner city, but they may at least contribute toward making some of those problems more amenable to solution.

Nor are the benefits of the public access channel limited to the inner city. It is no doubt true that the problems of the inner city are more deep-rooted and more pervasive, and that the inner city peculiarly lacks orderly means by which to articulate them. But one would be hard put to find any community—the bedroom suburb, the middle-class urban neighborhood, the low-rental area where the newly married are most likely to set up housekeeping—that does not have its internal communications problems, or an urge for cohesiveness that is not met by existing media.

Problems that Open Access Creates

The problems involved in the provision and operation of public access channels are admittedly enormous. The relationship between a television system and the citizen is a complicated one, and must be resolved by institutional arrangements of some kind or another. This in itself is reason for encouraging control, by preferential franchising rules, if necessary, within the community of a cable system serving that community, particularly in areas which are conscious of profound differences between themselves and neighboring communities. Such control might be lodged in a non-profit organization, or a profit-making corporate activity; in either case, there would certainly be a larger chance that it would be in tune with the community and reflect the community's desires.

The problems of access are still more formidable. Even cable television cannot cope with everyone shouting at once; there must be allotment of time, and a procedure for sharing out the most favorable times. There are problems with those who wish to voice highly unpopular opinions, and far sharper problems when the opinion can be interpreted as incitement to riot, or sedition, or defamation of character.

Public access channels, in short, are not likely to operate smoothly. But if they can help contribute in any significant way to the solution of the general problems within their communities, the problems they themselves create will be more than tolerable. The Commission recommends, therefore, that public access channels be made an essential element in cable franchises that may be awarded in a large metropolitan area; we believe further that consideration should be given to their inclusion in any other area for which a franchise is sought.

Rate regulation for the public access channels may also become, in time, a matter at issue. It is clear that the phrase "public access" is in practice meaningless if that access must be purchased at rates beyond the means of the general public within the community in question. As long as cable systems maintain excess capacity—and such a condition will certainly obtain over the next few years—economic considerations alone will keep public access prices at their lowest level short of direct subsidization. While the cable operator can only benefit, on balance, from additional diversity of programming, self-interest will keep rates at a minimum. It is true that upon occasions non-economic interests might lead an operator to use his rate structure, for reasons of his own, to restrict access; this is a matter upon which the franchising authority should certainly retain the power to exercise oversight.

As demand begins to challenge capacity, one might expect to see a general rise in channel rates, in which the public access channels would inevitably be caught up; if public access rates remain low, as they must if the channels are to serve their purpose, at some point they will appear to constitute a highly uneconomic use of channel space. That is not merely a problem of the public access channels, however (although it might affect those channels first)—rather, it is a problem of the orderly growth of the cable system, and the degree to which that growth should be governed by regulation.

Promoting Access

The provision of channels, even at highly favorable rates, will not be sufficient to bring public access television into use. Without a promotional force within the community, capable of providing technical assistance to groups who wish to use the channels, they will not flourish.

A study of public access in New York City, where such channels are mandated by franchise agreement, will be found in Appendix C of this Report. It indicates the difficult tasks that must be confronted by any promotional agency dealing with public access channels. They include the monitoring of cable management of access, representation of the public in the formulation of rates and regulations, educating community groups in the manner in which they can use access to further their purposes, assuring the existence of low-cost production facilities, and furnishing seed money and training for actual production. Where such a promotional force is absent, the public access channels are most likely to lie fallow.

The nature of that promotional force is likely to vary with communities. In those instances in which the cable installation itself is community-owned and community-controlled, the operating entity itself will hasten to provide the promotional efforts which may be required: that, presumably, is one of the reasons for community control. In the short term, while excess capacity is a burden on the installation, the self-interest of the private operator may lead him, for a time at least, down the same road. In other instances, purposeful intervention by appropriate governmental agencies—the Human Resources Administration in New York City, for example—may be necessary. Where high levels of community involvement already exist, the process may be self-starting.

An important role might and should be played by Public Television. Stations within the Public Television system possess technical knowledge and production facilities; the better stations are already in close touch with neighborhoods within their signal area. In Boston, WGBH has made its facilities available to community groups that wish to broadcast over Channel 44, its UHF community-oriented outlet. In New York City, WNET has cooperated with groups wishing to use the city's public access channel, making available space, production personnel, and production skills.

In recommending, as we have, a new orientation of Public Television toward the provision of local news and an involvement in community needs, this Commission may be charged with moving beyond its own terms of reference. We feel we are justified in that the combination of the skills and the devotion of Public Television with the capacities of cable television is such that there can be created out of those components local services that have heretofore been far beyond the reach of the television system. We are made more confident in this view by a communication from John W. Macy, Jr., president of the Corporation for Public Broadcasting, in which he proposes very much this role for the Public Television system.

The Traffic Director

There is no escaping, on cable television or on any other kind of television, the complications of the time-dimension. It affects television in many ways. The hour between 8 P.M. and 9 P.M. on television is by no means the same as the hour between 3 A.M. and 4 A.M. Neither is the value of one hour at the same time each week the same as the value of a scattered fifty-two hours during the year, even if they are scattered within the same general time period. If an event gains in urgency when it is transmitted live, no allotment of time after the event has taken place will serve the purposes of the producer. The production that demands several hours of sustained attention of its audience cannot be presented in segments, however favorable the time in which those segments are made available.

Thus there must be flexibility in the allotment of time by whoever it may be that makes that allotment. Scheduling must recognize the diverse categories of users, each with special time needs. Applicants should not be obliged to queue up for each hour they wish to reserve; if groups are to be encouraged to plan for the use of public channels and to set aside funds for the purpose, some regularity of scheduling must be assured. Certain periods might be reserved for brief items at reduced rates: the equivalent of the classified advertisement. In the light of the varying nature of the demands that may be made on the channels, to assert merely that allotment will be non-discriminatory, or on a first-come-first-served basis, is in practice to say nothing at all. There will be decisions that must be made, and consequently someone charged with making them.

Because the utilization of the public channels will begin slowly, the Commission believes that there will be time to experiment with a variety of approaches to the solution of the problem of the "traffic director," within a variety of installations. It could be the cable owner himself, although most will shun this role; if cable ownership is sufficiently diverse, this in itself will be an experiment. Our own preferences lead in other directions: we would prefer some such instrument as a public-access commission with representation from civic and cultural groups, or existing civic institutions within a community which assume this task as an additional responsibility.

Abuses of Public Access

Opening a camera lens and a microphone to anyone who may come by involves risks. Any actions designed to minimize or obviate the risks can be interpreted as censorship, whether they take the form of cutting the live communication off the air, refusing to transmit the tape or film or denying future access. This defines the dilemma with which television must live, and it is particularly acute—or at least particularly feared—with respect to public access channels.

Three of these risks are historically bound up with the regulation of radiated television. The risk that opinions will be stated without a corresponding expression of counter-opinion is covered by the fairness doctrine. The risk that there will be a biased presentation of partisan political issues of candidacies is covered by the equal time rule. The risk that a person will be subjected to personal attack, without the opportunity to defend himself, is governed by the right-to-reply rules.

It is the Commission view that the first two of these doctrines are not relevant to cable television, at least at the present time, and further that they are not likely at any time to be relevant to the public access channels. The manner in which the public access channels are established, the manner in which the Commission hopes to see them governed, and recommendations that the Commission will make in this Report to assure their copiousness, provide equal access for counter opinion and for the expression of competitive political views.

Right-to-reply rules, or some variation upon them, may be required. It seems unreasonable that a person who has been subjected to a personal attack which he did not invite, and over which he had no control, should be obliged to pay for the privilege of defending himself. In some instances he may possess, of course, recourse to the laws of libel, but although they may reward him for injury he has suffered they do not repair the injury. It should be recognized, however, that the invocation of right-to-reply rules implies the right of the cable operator, or his agent, to protect himself by cutting off, or by refusing to air, the original attack. In other words, there will be a kind of censorship at the source, which will at times be exercised by the engineer in control of the transmission. There are many instrumentalities by which the actions of the engineer, or any other censor, may be assayed after the fact, and it is not likely to be surpassingly difficult to see that they will, in general, be exercised prudently; it is true also that personal attacks are easily identified and rarely misidentified, so that the task of justifiable censorship is not unduly arduous.*

Beyond those three special risks are the general risks of criminal libel, incitement to riot, obscenity, sedition and the like. Any one of those offenses may be committed before an open camera, or recorded on tape or film. It is in these areas that there is little disposition to trust the judgment of the individual acting as censor: one man's obscenity may be another man's blunt expression; one man's incitement to riot another man's assertion of a political point of view. Our own society prefers that these issues be left to the judgment of judge and jury, and that offenders be punished *ex post facto* when it has been determined that a crime has indeed been committed.

In conventional television, it has been presumed that the operator of the television station is in the first instance responsible for the commission of any of those crimes over his station; he is thus automatically endowed with the right to censor, in his own protection, and he is likely to make full exercise of that right. It is clear that such a presumption is unwarranted with respect to public access channels (and to other leased channels) and inimical to the most useful operation of those channels.

The Commission therefore recommends that the appropriate legislative bodies modify existing laws to remove the liability of the cable operator in these respects so far as channels outside his direct control are concerned. The burden of meeting the requirements of the civil and the criminal laws will thereupon fall entirely upon those who brought the burden upon themselves.

In extreme cases, the operator or his agent will no doubt terminate a transmission or simply refuse to transmit a re-

---

*One member of the Commission, Mrs. Wald, dissents from this general position. Mrs. Wald holds that a cable operator or his agent should not be allowed to interfere with statements because he fears that those statements might provoke invocation of the right-to-reply rules.

corded telecast; certainly no cable operator will allow what might be called "hard-core obscenity"—of which the definition now appears to alter from year to year—to pass over his system. If a community or an individual should take the view that the privilege is being abused, it appears to this Commission that there is a variety of remedies to which the community or the individual may have recourse. In practice, it is not an outcome that is likely to be frequent. The repeated transmission of material that the community in general finds offensive does not in fact appear to be a significant threat. The experience of Channel 44 in Boston is instructive: it stipulates rules that participants in its public access experiments must abide by, and has found that this procedure keeps offensive material at an extremely low level.

On the whole, the Commission is not as disturbed as some in the face of all these risks. It appears to the Commission that they represent situations that will not often occur, and that can be readily managed on those rare instances when they do occur, as they are managed with respect to other media of communication. The greater risk, we believe, would be to impose regulations that would stultify the public access channels.

With this discussion, this Report concludes its account of the configuration that it expects cable television to assume, sometime within the present decade. It will carry mass entertainment, some of it picked up from conventional television, some procured directly by cable television or its lessees, most of it supported from advertising revenues. It will carry special entertainment, supported by subscription, by advertising revenues, or by a combination of the two. It will transmit news and opinion, supported again by some mix of advertising and subscription revenues, but in all likelihood heavily dependent on the latter. It will possess channels dedicated to governmental and quasi-governmental services, and others (with which this Report has not dealt in detail) to educational services and to commercial services. It will possess, finally, mandated channels for the access of the general public and institutions representing the general public.

How all this is to be instituted and managed, and how it is all to be fairly accommodated within the limits of the cable system, remains to be considered.

CHAPTER TWELVE

OWNERSHIP AND CONTROL OF CABLE SYSTEMS

An instrument of communication is an instrument of social power. Even when its primary use is for the provision of entertainment, it conveys as well, subtly or not so subtly, value judgments and opinion. When news and commentary are a part of the message, or all of it, communication is explicitly intended to affect what the recipient thinks and how he behaves; it may not be sufficient to determine his response to the world in which he lives, but certainly it colors that response.

There is consequently a general repugnance to the notion of any kind of monopoly over instruments of communication. One would prefer a situation in which many voices may be heard, and in which the citizen may have the opportunity to select the voice to which he will give credence.

Licensing power over radio, a necessity from the outset if there was not to be chaos on the airwaves, brought with it an opportunity for the government to help assure a multiplicity of voices. That opportunity was seized by the federal government which limited the number of radio stations that one individual or one corporate enterprise could own. This was what might be called a horizontal restriction, concerned with the country as a whole. There was also possible, even in those early days, a vertical restriction, which would have prohibited undue concentration under one ownership of many different media within a given geographic area: regulations, for example, that would have prohibited a newspaper proprietor from owning a radio station in the same city. At the time, no need for vertical restriction was felt. The citizen, in most urban areas, had a choice among many newspapers and many radio stations — enough of a choice to provide all the voices he would be likely to heed.

In the intervening years, circumstances have altered. Newspapers have merged or failed; there are only a handful of cities with as many as three under independent ownership, and few with as many as two. The restrictions upon the number of television stations, in any single area, are far more severe than the restrictions on the number of radio stations. A competing television voice, unlike a competing radio voice, cannot be picked up from a distant source. National mass magazines have dwindled.

Dangers of Cross-Ownership

Where media compete for advertising revenues, moreover, cross-ownership — the common ownership of nominally competitive media — creates the opportunity for economic manipulation. The advertiser finds one vendor bidding for his business where previously he might have enjoyed the advantages of trading off one against another. The proprietor can arbitrarily favor one of his enterprises over the other, and the advantages of open competition between press and television, for example, can thus be lost.

In recent years, the FCC has become progressively more sensitive to these issues. Nonetheless, television retains close links with other media. In 1968, 127 of the 184 television stations in the fifty largest markets were owned in common with another communications medium, which was often the only newspaper enterprise in the central city.

The appearance of cable television, and its potential for growth to the size and scope postulated in this Report, does not alleviate the problem and in some degree even aggravates it. In a city which possesses three television stations, one of which is owned by the sole newspaper enterprise, there are in effect three separate voices. If a single cable system within that city is owned by any one of the three stations, there are still only three voices. But one of those voices has suddenly become far more powerful than either of the other two. It possesses control over twenty channels or more of communications, where a television station has only one, or a television station-newspaper complex has only two or three. The degree to which communications power, and economic power, are lodged in a single hand is thereby intensified.

There is a further complication arising out of cable television's threat to the status quo. An over-the-air television station in a large city may be an extremely profitable enterprise. The appearance of cable, to the owner of such a station, can be seen as a threat with unpredictable consequences; his self interest may appear to him to lie in seizing control of the cable franchise and using that control to prevent development of the franchise. Since we believe the general interest is served by encouraging the growth of cable, we are obliged to regard such an outcome as undesirable.

Merits of Cross-Ownership

But despite these arguments against cross-ownership of cable television and other media of communications, there are arguments in favor of cross-ownership as well.

—It can be argued that the participation of newspapers and

local television stations in cable television can contribute significantly to the capacity of cable to serve the community. Those who are engaged in communications within the community bring to cable talents and skills which are essential to the services cable should provide. In particular, the association of a local newspaper with a cable system may be the most efficient means of assuring that the cable system will assume responsibility for local news coverage.

— Capital will be needed to make cable grow; in a sense the price of the cable system is raised and its quality diminished if a significant sector of potential entrepreneurs is excluded.

— Although ownership by a newspaper will not add to the diversity of voices, it may help preserve a diversity of modes of communications to the extent that ownership of a cable system may enable a newspaper to remain in operation that otherwise might fail. That has sometimes been one of the consequences of cross-ownership of newspapers and conventional television.

— Some of the threat of a monopoly effected by cross-ownership rests on the unstated assumption that only a single cable franchise will be issued in a given marketing area. In practice, there are likely to be several franchises in large cities, and even more when entire metropolitan areas are considered; regulations or franchising provisions could readily be devised that would assure diverse ownership of the several franchises, and hence diverse voices.

— As a practical matter any steps to bar television stations and newspapers from ownership of cable franchises would have the effect of calling into being powerful opposition to cable television in general. The effect of that opposition and the support it could muster both nationally and locally, might be a greater threat to the healthy growth of cable television than the fact of cross-ownership itself.

— There are arguments of equity in favor of permitting those interests most threatened by the growth of cable television to balance the threat of economic loss with an opportunity for economic gain. It would not have been considered equitable, some sixty or seventy years ago, to have prohibited buggy makers from trying their fortunes in the business of building automobile bodies. It can be argued that neither is it equitable to prohibit broadcasters from trying their fortunes in cable.

—Finally, many of the arguments against cross-ownership equate ownership of cable franchises with control over cable channels. It is possible to limit that control by mandating certain uses for some of the channels and by obliging the franchise-holder to offer other channels for lease; this Report offers recommendations to this effect. Much of the monopolistic threat of cross-ownership is thereby diminished sharply.

A Compromise Position

We have, as a Commission, attempted to work our way through this maze of arguments and counter-arguments. Our conclusions appear to us to be the best and most workable compromise between opposing positions and are those that we believe are most likely to encourage the healthy growth of cable television.

In seeking a compromise position, we found the economics of cable television in our favor. So far as cable franchises are concerned, economies of scale dwindle after a few tens of thousands of households have been reached by any single system. This fact signifies that from the standpoint of economics, at least, any area that supports a commercial television station or a local newspaper will be able to support several cable franchises, each of them well above the minimum economic size. The very densely populated area, which supports several television stations and two or more newspapers, will provide opportunities for many cable franchises, all quite sizable.

Thus we are not necessarily faced with an all-or-nothing

situation in regard to cross-ownership. It is possible to award cable franchises, within any sizable metropolitan area, to one or more television stations and newspapers without thereby conferring dominance of cable television upon competing media.

Our proposal, therefore, rests upon the concept of the Standard Metropolitan Statistical Area, as defined by the federal government. We recommend that within any SMSA any commercial station or newspaper of general circulation be permitted to seek a cable franchise or group of franchises capable of reaching no more than 10 percent of the households within that area. We recommend further that in no SMSA should ownership by commercial stations and newspapers be permitted to exceed, in the aggregate, 40 percent of all households.[1]

These recommendations, taken together, permit entry of competing media into cable television within large market areas, without the risk that they will dominate cable television either substantively or economically. So far as ownership of cable franchises outside their own markets is concerned, we believe no prohibition should be established. In addition newspapers would, of course, be permitted to lease channels on cable systems within their own market areas or anywhere else.

We recommend that network ownership of cable franchises be prohibited. We believe that there may be a clear conflict of interest between existing network and cable operations, which might in the present respective strengths of the two industries inhibit the growth of cable television if they were under common ownership. This recommendation reflects the view of the majority of the Commission; a minority believes network ownership should be permitted.

In the interest of preventing horizontal monoply, we recommend that a limit be placed upon the number of cable subscribers served nationally by any single individual or enterprise.

[1] One member of the Commission, Mrs. Wald, qualifies her concurrence with the reservation that the figure of 40 percent, in the aggregate, should be substantially reduced. Another member of the Commission, Mr. Thomas, did not participate in the final discussion of this recommendation, and is not recorded.

That limit might be set at 10 percent of all households within the reach of cable at any given time.

We recommend that Public Television stations be permitted to seek cable franchises without market area restrictions. We see no threat of media domination, economically or substantively, by Public Television.

We recommend, finally, that all cross-ownership restrictions on local commercial stations and on local newspapers be removed if no other bidders for a franchise appear.

We do not believe there is justification for restrictions on publishers other than newspaper publishers, or on operators of radio stations.

It appears to us that the only other kind of regulation that might be put forward by other bodies would be *ad hoc* regulation, in which the circumstances peculiar to any given area would govern regulation in that area. Each area would thus present its separate, if not unique, problem. We believe such a procedure would be unworkable, either at the municipal level, where political pressures are great and the public interest not always best served, or at the federal level, where the complexity of the problem would be staggering and where the likelihood would be minimal that consequent error could be corrected.

Mandated Channels

The distinction between ownership of a cable franchise and control over the use of channels, to which we have alluded, is significant. We believe that the franchising authority should possess the power to mandate, within limits, the uses to which some proportion of the channel capacity is to be applied, and that in addition the federal regulating authority should promulgate rules to assure that a proportion of the channels be available, on a first-come, first-served basis, for leasing.

The mandated channels fall into two categories: the channels which must be set aside to carry network and local independent stations, and the channels intended to provide room for the range of services enumerated (in part) in Chapter Nine of this Report: health, welfare, consumer services and the like, and to provide public access to the system, as described in Chapter Eleven.

In a typical major metropolitan area, we believe these mandated channels should occupy approximately one-half the capacity of a twenty-channel system. Of the twenty channels, a maximum of six would normally be mandated for network signals and local broadcasting stations; two for service uses (other than education); one for public access; and one for experimental educational uses. Of the two channels set aside for service uses, one might reasonably be allocated to the government, the other set aside, if a demand exists, for quasi-governmental institutions.

More or less the same proportions should be maintained in larger systems. A forty-channel system would have room for ten network and local stations (and indeed would require that capacity in some cities); there would then remain four for service uses, of which two or three might go directly to local and state government; two for public access; and two for experimental educational uses.

All of these figures should be construed as minimums. The franchising authority should, if it so desires, require of the franchise holder more mandated channels in any category or make provision of mandated channels above the minimum a bargaining point in the issuance of franchises.

Leased Channels

Of the remaining channels, we recommend that the operator be obliged to set aside all but two for leasing on a first-come, first-served basis. He would himself be entitled to operate two channels, and any channels available for leasing for which no bidder appeared. The types of arrangements that may eventuate, both with respect to first-run and subsequent-run programs, are described in Chapters Five and Six, which deal with those matters. But the leased channels will, by definition, be available for more than entertainment and news. They will be available to any bidder, profit or non-profit: to the government agency which wishes to augment the services it provides on its mandated channels; to educational institutions, private and public; to those interested in the political process.

Out of this combination of owner-operated channels, mandated channels, and leased channels will come, we believe, the diversity of voice and the diversity of presentation that we see as the greatest contribution cable television can make to electronic communications. To render the process even more efficient, we would add one further restriction: a limit should be imposed upon the number of channels any single lessee may operate on any single cable franchise. Such a restriction is necessary to assure that the maximum diversity is maintained. There are already proposals that are intended to permit a single programming agency to dominate as many as eleven program channels, and to do so on a national scale. Limiting any lessee to a maximum of two channels would eliminate the threat. The prohibition might usefully take effect only when cable penetration had passed the level of perhaps 30 percent in the area under consideration.

If cable television grows as we believe it may grow, almost inevitably franchises of smaller channel capacity will begin to experience symptoms of scarcity with respect to leased channels. The market would normally lead them to respond by raising the price of leased channels—a course of action which might certainly be in their short-term interest, but would not necessarily be in the interest of the municipality involved or of the public. Surely there is little reason why a radiated television of scarcity should be replaced, admittedly on a higher level, with a cable television of scarcity. We recommend that franchises include provisions that will enable the franchising authority to demand of the operator, upon evidence that his channels are in demand substantially beyond their capacity, that he increase channel capacity to conform with new and predictable needs.

We believe also that there should be flexibility with respect to mandated channels. The public is not served if the government, for example, or the educational system, possess rights to the use of a channel which it does not exercise: a vacant channel is a resource going to waste. We recommend, therefore, that the federal government be prepared, after perhaps five years of experience with mandated channels in areas of high cable penetration, to reexamine the question of mandated channels and the proportions that are dedicated to each use, with a view to remedying imperfections in the general system that such a study might reveal.

Regulation of Rates

The monopoly position of the cable operator raises questions of rate regulation. Because the system owner will be leasing channels and selling subscriptions, it may appear that he will be in a position to exact monopoly profits at the expense of the subscriber and of the channel lessee.

So far as the lessee is concerned, it appears to the Commission that the marketplace should be an acceptable regulator of the price of the lease, particularly if there are provisions in the franchise which require the operator to increase capacity to reflect increased demand. The marketplace will mediate the demand for channels far more satisfactorily than any system in which rates were artificially kept at a low level.

This should not be taken to preclude the possibility of preferential rates for special users. Such rates, set perhaps at a fixed percentage of going commercial rates, might apply to educational institutions and certain non-profit organizations operating in the public interest.

As the income increases that a cable operator derives from his rented channels and from the channels he programs himself, he may find it to his economic advantage to eliminate the basic monthly subscription fee. That is, he may decide that his revenues would be greater if he removed the disincentive represented by the monthly fee, and thereby built his subscriber list faster than he otherwise could. Television might once more revert to universal provision of free basic entertainment and hence universal access to television's services and to pay television. We do not count upon this, but neither do we rule it out.

In the meantime, however—and "the meantime" is in this instance likely to be indefinitely long—the marketplace cannot regulate the price of the basic subscription fee, since in this respect the monopoly position of the cable operator means that there will be no direct market.

Internal Subsidization

Franchising procedures generally include the imposition of maximum rate regulations; we believe they should continue to do so. We question the wisdom, however, of franchising practices which are based wholly or primarily upon rates. The franchise should be awarded not to that applicant who promises the lowest rate, but to that applicant who promises the best combination of low rates and high services.

A promising franchising procedure already in use merits widespread attention. In the first stage of the process, applicants are asked to specify standards and services and the rate at which they would be provided. The franchising authority, on the basis of its study of those initial bids, then determines the package of technical and service requirements which an acceptable application should offer to provide, and asks for second bids, this time on the basis of that package. The low bidder on the second round wins the franchise.

It is clear that such a procedure, or one which would achieve the same ends, calls for the applicant to provide "internal subsidization" of some services by the more profitable segments of its operation. Among those services, at the outset, might be the commitment to charge non-discriminatory subscription fees, to bring cable to some parts of the area which might not be profitable, to provide at low cost head-end facilities for public-access channels, and to install capacity and digital return capability which will initially be far beyond the needs of the system.

If support for public services is to be sought from cable television, the encouragement of such internal subsidization appears to this Commission preferable to any attempt to raise public funds directly from cable television, whether by direct tax or by indirectly siphoning off funds. Taxes intended to cover municipal costs in monitoring the system are of course reasonable; any tax beyond that, other than ordinary personal and business taxes, would be initially regressive in nature. In particular, a gross revenues tax will have the specific effect of prejudicing the growth of cable systems, and is firmly opposed by the Commission.

Common Carrier Status

Under the system we recommend, taken as a whole, there would be two classes of distributors of programs. By far the more important class, initially at least, would be holders of FCC station licenses, who would be providing programs by way of the channels mandated to carry network and local signals, and over the air as well. The newer class of distributors would be the lessees of cable channels and, if he so chose, the cable operator himself.

The restrictions on the first of these two classes are in some respects moderately severe. Station licenses are renewed every three years; in principal, renewal has up to now been based on the quality of past service. The holder of the license feels obliged in some degree to meet FCC programming standards, to provide public services, and to operate in the public interest; the temporary nature of his license, in addition, makes him sensitive to pressure from public groups, such as groups recently formed to alter the nature of children's programming. He is subject to reporting requirements and to the operations of the fairness doctrine and the equal time rule, when either applies. The network affiliate is limited to three hours of prime-time network programming. The station must assess community needs and report to the FCC on its efforts.

Lessees and cable operators are under none of these obligations; nor are we now proposing that they should be. Yet if it should turn out that one cable channel commands a substantial part of the national audience, comparable to the audience now commanded by a major network, a reevaluation may be called for. It may be found advisable either to remove existing restrictions on stations or impose them upon cable installations.

There have been many who have suggested that cable television should be regulated as a common carrier and the owner of the system prohibited from operating channels or controlling programming in any way. The objective of such a system would be to encourage the maximum amount of diversity and untrammelled access to the cable without fear of inhibition by the system owner, who might be tempted to discourage programming in competition with his own or programming that he feared might offend audiences and so limit his subscribers. The arguments in favor of divorcing ownership from transmission derive additional strength from the inherent monopoly of a cable system in any locality as opposed to the competition of several over-the-air broadcasters in most places.

Common carrier status may, indeed, be the way cable should and will go as it achieves maximum penetration and overtakes or supplants over-the-air broadcasting. At this point in time, however, the Commission believes that imposition of common carrier status would be unrealistic and an impediment to the desirable growth of cable. We do not believe that investors would be willing to undertake the substantial capital expenditures of laying cable if they had no control over the use of the channels in the formative years and so were powerless to control the financial destiny of the system.

We have tried to postulate a system which can be converted at a later date to common carrier status, if that proves desirable, but which initially will be attractive to investors. At the same time, our recommendations of required reservation of channels for nondiscriminatory public access, for service uses, and for compulsory leasing should insure to a large degree diversity and widespread access by a myriad of voices not subject to the control of the cable system owner. As cable grows, the community may wish to increase these requirements and limit the owner's stake in programming to the point of total divorce. We prefer to leave that decision to the future. Our recommendation for relatively short duration of franchises would permit such an option to be exercised at periodic intervals.

Diversity of Ownership

The general intent of these recommendations on ownership and control is apparent. We believe that cable systems and cable channels ought not to be developed, owned or managed by the same people, or even by people representing the same set of interests, in every community across the nation. The product cable can sell is not like ingots or heating oil, of objectively measureable quality, and desirable at the lowest price and greatest quantity. There is much that is unknown about the potential of cable television, and much that will never be learned —things that will not happen—if its development is limited to a small group of entrepreneurs or a common type of entrepreneur.

The public interest, certainly in this period of growth and experiment, will be best served if cable systems are owned, and cable channels managed, by a wide range of interests, including non-profit community groups and organizations representing minorities within the larger public. Because cable systems, through their current operations, will inevitably make many of the rules and determine programming patterns, at least for the vital next decade, a diversity of system owners and channel managers will contribute significantly to an imaginative testing of the medium's possibilities. We recommend that local communities affirmatively seek to insure diversity in ownership and management, and recommend further that communities actively seek to promote this end by permitting franchise areas to coincide, where that is practical, with sub-districts within the city.

Because the whole enterprise of cable television will be so much in motion during the next decade, and passing through so rapid a learning period, we recommend that municipalities limit the duration of franchises. Franchises must be long enough to attract capital. But very long franchises perpetuate arrangements and ownership patterns that might seem appropriate today but that experience and further information might show to be unsatisfactory or dead wrong. We recommend that no franchises for a term longer than ten years be awarded, but that franchise terms include provisions for purchase at a fair price of the assets of an unrenewed franchise.

We do not expect universal assent to all these recommendations governing ownership and control of cable franchises. We believe them, however, to be reasonable and equitable, and we believe further that they are such as to encourage the growth of cable television over the next decade in a fashion that will best serve the public.

CHAPTER THIRTEEN

THE REGULATION OF CABLE

Preceding chapters have described the kind of regulatory powers we believe should be exercised over cable television, and in rough terms the scope of those powers. They are both fewer and more narrow than the powers exercised over conventional television, and we have not been anxious to seek out areas in which government might play a larger role. Nonetheless, the regulation we favor in this Report is extensive, ranging from matters of ownership and control of cable systems and cable channels to the relationships among cable television, conventional television and program supply.

In this chapter we are concerned with the level of government—federal, state, or local—at which the power should be exercised. The decision on locus of regulation is quite as important as any decision concerning the content of the rules themselves; both will have important consequences for the growth and structure of the industry.

The Ineffectual Past

It is difficult to assign responsibilities; the past performance of regulatory agencies, at all levels of government, has been less than laudable. In the first two decades of cable growth, the federal government has been rudderless, the municipalities inept, and the states inactive.

In an earlier chapter we traced the shifting course of FCC attitudes toward cable: from indifference to hostility to mild encouragement. The FCC, for its part, can reasonably protest that it has been given no useful guidance by the Congress. Although cable presents problems of fundamental national communications policy, Congress has made no serious move to deal with them. The Communications Act of 1934, as its date would imply, is not explicit on the scope of FCC authority over cable, and Congress has yet to clarify the areas of ambiguity. It has yet to enact legislation on the difficult copyright issue. Congress has not been entirely passive; particular committees are not reluctant to challenge the FCC when it takes action they dislike. But substantive action of their own has been lacking.

Cities, by default, have seized upon the necessity for cable television to use public ways as justification for the application of franchising procedures, and more often than not have looked upon the entire process as one of assuring new revenues for the municipality. This in itself is scarcely the most satisfactory attitude, and since in these early days of cable television the revenues are themselves almost insignificantly small it does not lead to any serious approach to the problems of cable. But there are also deep institutional reasons for the poor showing of local government. Cities have no experience in regulating a communications industry: the franchising authority, if there is one, is more accustomed to deal with buses and taxicabs. Nor do they possess adequate machinery for the enforcement of franchise conditions.

Our study of the franchising process has revealed that city officials often make brave attempts, but do not in fact have the capacity to evaluate the offers of cable operators in the light of community needs; that franchises are usually awarded, especially in smaller communities, after only cursory discussion; that at times franchises are awarded and the systems never built; that some systems are so constructed as to skim the cream of the business; that franchise terms are often imprecise; and that there is great difficulty in exacting compliance, so that even if the system is in fact built it does not deliver what the operator promised.

Only lately have states, a third source of regulation, exhibited interest in cable. So far, seven states (Connecticut, Vermont, Rhode Island, Nevada, Illinois, Alaska, and Hawaii) have sought to regulate cable. New York and New Jersey have imposed moratoria on continued municipal franchising, thereby recognizing that cable growth presents issues of statewide significance whose complexity might exceed the capacity of local officials to deal with it. Other states, among them California, Massachusetts and Iowa, have launched inquiries intended to define an appropriate state role.

Regulation Where Necessary

The sum of all this regulatory chaos has been a long period during which healthy growth has been all but impossible. No party to the subject is content with matters as they now stand. There are divisions within the FCC, and factions within Congress. Municipal governments, without clear guidelines, fumble for solutions. States have for the most part stood clear, but when they do enter the arena are most likely to look upon cable television as another kind of telephone service and to regulate it accordingly. Clarity can come only if legislative action is taken, at both the federal and the state level.

Looking at the problem in the large, the demarcation between federal and local jurisdiction can readily be drawn. The statute creating the FCC directs it to "make available, so far as possible, to all the people of the United States a rapid, efficient, Nation-wide and world-wide wire and radio communication service with adequate facilities at reasonable charges." Cable television was not anticipated in the statute, but its general intent applies as well to the new technology as to the old. To the extent that cable television is a national system, it calls for regulation—where regulation is necessary—at the national level.

But cable television differs from radio in its heyday, and from present-day conventional television, in that much of it is an intensely local activity. It does not depend upon transmitters broadcasting a signal over tens of thousands of square miles. Much of its channel use (although not necessarily the same proportion of its audience) will be at the municipal and even the neighborhood level. If it is to be fully responsive to local needs, there must be a regulatory power not too distant from those needs to sense their character. To the extent that cable television is a local system it calls for regulation—where regulation is necessary—at something closer to the local level.

But these are broad statements. To give them real meaning, their implications must be worked out.

The Federal Level

It is beyond the scope of this Report to specify in detail federal legislation and regulation with respect to cable. We content ourselves with specifying some of the more important features such action should embody. The premise must be that cable will be, in time, not merely a supplement to over-the-air television but will evolve into a principal means for the distribution of entertainment, news and other services to the American public.

— Congress should resolve the copyright issue. We have recommended that it do so by making programs available for purchase by cable television on terms commensurate with those applicable to broadcast television. Only Congress can enact this legislation, by bringing up to date its statutes on the nature of property rights in intellectual products.

— Legislation should acknowledge the power of the federal government to protect the viewing public if the growth of cable threatens to deprive any significant sector of the public of broadcast service.

— Congress or the FCC should set minimum standards for the allocation of channel uses within local cable systems, as among owner-operated channels, channels available for lease, channels set aside for services, and channels set aside for public access. We have made appropriate recommendations in Chapter Twelve.

— The FCC should establish minimum technical standards for all local cable systems, to the extent that standards are necessary to assure interconnected operations, and to assure as far as possible that material prepared for transmission on one local system will not be excluded for purely technical reasons from another.

— Congress or the FCC should require that all new local cable systems provide channel capacity sufficient to ensure a fully national system.

— Congress or the FCC should require that access to all leased channels and public access channels be non-discriminatory. Any aggrieved person should be enabled to seek immediate relief.

— Congress or the FCC should assert limitations on ownership and on leasing that will prevent undue concentration of power over the medium, or over media of communications in general, and that would deny ownership or channel control to those for whom it would pose a conflict of interest. We have made appropriate recommendations in Chapter Twelve, some of which coincide with existing rules.

— Congress should limit the powers of states and municipalities to impose a gross receipts tax or similar levy on cable systems. There are First Amendment considerations involved in restraining the zeal of municipalities to exploit the communications media as a major source of revenue.

We believe that a tax which yields sufficient revenue to cover the cost of regulation is appropriate.

Legislation in those categories would provide, we believe, a reasonable basis upon which a national cable television system would be likely to develop. But we also see need for a quite different kind of legislation, directed not toward the regulation of cable television but toward its most effective utilization.

A Federal Promotional Agency

We believe Congress should authorize and provide funds for a program that would encourage the wise use of cable. Promotional activities of that sort should, of course, be sharply separated from regulatory activities; they might be assigned within the Department of Health, Education and Welfare, which has dealt in the past with matters concerning the content of television and which has substantive concerns of significance to cable television. The creative use of television for purposes other than entertainment and news is the great promise of cable. To bring it into being will require effort, leadership and money, which we believe it appropriate for the federal government to help provide.

Such a promotional agency might function in the following fashion:

— As a national clearinghouse for information concerning the uses, prospects, regulation and promotion of cable;
— To evaluate the development and uses of cable in the various states and to advise state and local governments on the development and uses of cable;
— To develop a federal plan for the financing of experimental and social uses of cable for which commercial support is not obtainable;
— To make recommendations to the Congress for new legisla-

tion concerning the further development of cable and its use in the service of society;

— To stimulate the creation of corresponding state agencies, and perhaps to make grants of funds to those agencies; the latter would, indeed, be the best means of stimulation.

Cable television, even at its present level of growth, is large enough to warrant this kind of attention from the federal government. In the immediate future, it will take the shape outlined in this Report only if it receives that attention.

The State Level

Problems at the state level arise out of the mismatch between the technology of cable television and the boundaries of local government, the absence among many local governments of the special knowledge that may be required to conduct and monitor the franchising process, and the special nature of local needs.

The various states include, as direct instruments of the state, county governments with various degrees of power, and incorporated and unincorporated municipalities ranging in size from many millions to a few thousands or even hundreds. There is no lack of governmental boundaries within the states; they come in all forms and enclose all kinds of constituencies.

To the operator seeking a franchise, the scene may appear to be one of total confusion. Even in the very large cities, where local government is strong, he may discover that his natural franchise area crosses out of the city into a suburb. Other areas may be such that no franchise enclosed within a single governmental unit is economically viable; the franchise applicant is obliged to make concurrent arrangements with several governing bodies. Elsewhere, where county, township and village governments overlap, it may be anything but clear to whom he should apply for a franchise.

We have alluded above to the special problems that face any municipality, except perhaps the very largest, when it comes to deal with cable television. Experience is lacking in the very elements of the issues that arise when a franchise is to be awarded. It is lacking, at present, among most of the states as well, but the states are in a position to organize in order to take advantage of what experience there may be. The state, moreover, being engaged in the problem repeatedly, will be able to amass experience of its own.

The state, finally, may be expected to have some awareness of the special circumstances that exist within its boundaries. We have recommended that the federal government establish minimum standards for the allocation of channel use, but those minimums may take on quite distinct aspects in Wyoming and New Jersey; where they might be quite sufficient in one, they may be quite insufficient in the other. New Jersey might well require a greater number of public access channels, a larger dedication to neighborhood service; Wyoming might seek standards that would assure state-wide service and feel far less need for services to distinct communities within the state. All of this should be within the power of the state to assure.

Similarly, a state might well wish to assume some direction over the manner in which the cable system takes shape. In New York, with special municipal needs, there might be an insistence within franchises that growth be distributed in some orderly fashion that would assure provision of service where service was most needed; in another state, the forces of the market might be permitted to operate freely.

The state, in short, should be in a position to set the terms for negotiation with a franchise-seeker in a fashion that will preserve local interests. Most states which have acted to that end have done so by placing responsibility for the regulation of cable television within their Public Utility Commission or an equivalent body. We do not believe that to be an appropriate assignment.

The Public Utility Commission deals primarily with the consequences of monopoly in the provision of public services. Their concern is rate regulation, return on investment, and the provision of uniform service. To some extent, these matters are also of concern in the regulation of cable television.

But there are other major concerns as well, as this Report has stressed: concerns that have to do with the differing nature of the service rather than its uniformity; with the problems of a great many small-sized or medium-sized monopolies existing side-by-side rather than one or a few large monopolies dominating an entire region. The problems are not at all the same, and it is inappropriate that an agency whose major attention is directed to one kind of regulation should be empowered as well to deal with another which may, in important respects, be profoundly different.

Special State Agencies

We recommend, therefore, that the states establish agencies empowered to direct and regulate the growth of cable television, in conformity with the standards established by the federal government but with freedom at all times to exceed those standards where they are expressed as minimal. The principal activity of such state agencies should be the general supervision of franchising procedures. In carrying out that activity, these agencies should:

— Identify appropriate franchise areas within the state, with due regard for existing government boundaries on the one hand, and for natural neighborhoods and communities on the other.
— Within those franchise areas, identify or bring into being

officially constituted franchising bodies to which applicants for franchises might repair.

—Establish, in its turn, minimum standards which must be
represented in all franchises. Those terms might cover
technical standards, allocation of channels, performance
standards, non-discriminatory access, duration of franchise, and the like. It would not involve itself in rate
regulation.

—Establish uniform accounting methods for cable operators.

—Require that rate-cards and franchise terms be deposited
with the agency, and that reports of compliance by
operators of local systems be submitted annually. All this
material should be made generally available.

—Receive and adjudicate appeals with respect to the performance of a cable operation.

The agency itself should not, except in extraordinary circumstances, issue franchises; that privilege should be retained
as close to the franchise area itself as it is practical.

These provisions, it might be noted, are in many respects
similar to those under which the states involve themselves in
education. Here, too, there is state regulation with attention to
the preservation of local autonomy; here, too, may be found
state assistance or intervention in the establishment of school
districts which may cross local governmental lines, and in the
assertion of minimum standards.

We believe that state agencies of this sort will diminish, if
they will not eliminate, the objections of cable operators to the
notion of state intervention. Their fear has been, in general,
that state regulation patterned after public-utility regulation will
inhibit investment in cable television by rigorous application
of rules concerning return on investment; that an essentially
speculative undertaking, at least while it is in its period of
growth, will be subject to rules more appropriate to huge and
irreplaceable major industries; and that regulation in general,

within an agency that has its mind on other matters, will be uninformed and arbitrary. The agency we recommend should escape these criticisms, although no doubt criticisms of another kind will arise.

In carrying out many of their functions, state agencies (and franchising authorities as well) will shortly be in a position to benefit from assistance made available by the Cable Information Service, now being established with support from the Ford Foundation and the John and Mary R. Markle Foundation. The Cable Information Service is being created with precisely that end in view; it is intended to serve as a neutral source of technical assistance and advice for all public agencies charged with responsibility for the development and regulation of cable television.

The City Level

There is ample precedent for the delegation of state power to the city, in its general form under terms of "Home Rule Charters," and often in detailed forms as well. Delegation of that kind reflects a recognition of the special problems of the large city, and its ability to muster sufficient resources to meet those problems.

We believe such delegation is of major importance with respect to cable television. The great metropolitan centers have their own necessities, and should be given the opportunity to meet them through their own resources and those that the Cable Information Service can bring to bear.

The goals of the franchising power have been expressed, directly or by implication, in the foregoing sections of this chapter. They do not differ between state and great city. The delegation of power to the city must be such that the goals are maintained. But we believe that the state must be prepared to let major cities set their own maximum rates and terms of performance; make their own channel allocations; establish their own appeal and adjudicatory mechanisms of first resort; and establish their own franchise boundaries where they are wholly within the city.

One special matter might be noted here with respect to the franchising power of the large city. We have noted on several occasions that the community enterprise, profit-making or nonprofit, is particularly appropriate to some of the broader purposes of cable television. We have especially in mind those enterprises which have come into being within neighborhoods that have special social or ethnic problems and constitute in some measure sub-cities with special requirements and special knowledge of their own.

We urge that wherever those circumstances are found, cities give preference to franchise applications from such enterprises. We do not propose that standards be waived; such enterprises should be required to meet technical and performance standards. But we believe that the best interests of the community, and hence of the city itself, will frequently be served if management of the local cable system is itself truly local, and representative of those who make up the neighborhood.

A Rationalized Division of Power

Cable has now reached the point in regulation that radio reached when the Radio Conference of 1927 was convened by President Hoover. The landscape is much the same: thousands of small operators, a few emerging giants, conflicting standards, and fruitless attempts at regulation at the state and local level. By 1927, the major entrants in the industry had come to believe that radio would not become a major social and economic force without regulation that recognized and confirmed radio's national role, that restricted competition, and that provided a climate of growth unencumbered by costly compliance with numerous, sometimes inconsistent state rules.

Today, in the cable industry, the major operators feel that any arrangement with the broadcasting industry defining the extent to which cable can carry over-the-air signals must be national in character, and that any settlement of the copyright controversy, if it is to be durable, must clarify the copyright liability of cable operators throughout the country. They know that if they are to obtain the large-scale financing that is necessary for rapid cable growth, piecemeal advances, however painfully secured, will not be sufficient. What the largest cable companies fear most is the uncertainty and delay that result where there is a multitude of regulatory authorities with inconsistent standards.

The division of power we recommend among federal, state and local authorities is one we believe will prove workable in practice. It reserves for local exercise the powers that are necessary to assure that the local cable television system meets as nearly as possible local needs; it preserves the customary division of authority between city and state and encourages the state to offer increasing assistance to smaller aggregates which must inevitably turn to the state for expert advice and guidance. Where national goals and purposes are in question, it looks to the federal government to provide direction and assistance.

Because we have sought to place minimum reliance upon regulation (although even that minimum is large) we do not believe cable television is likely to ossify under regulations hastily imposed, or imposed before the true shape of cable television has taken form. We have sought also the maximum flexibility in modifying regulatory practices in the light of experience, particularly during the decade directly ahead. We believe cable can grow, and grow healthily, under the conditions we have recommended. We look to Congress, the FCC and the states to perfect in the political arena, where they must be perfected, the suggestions we have put forward in this Report.

CHAPTER FOURTEEN

AFTERWORD

At the outset, the problem may not have appeared particularly difficult. Here was a new technology, not inordinately complicated, which made it possible to replace the insubstantial, somewhat mysterious electro-magnetic wave and all its shortcomings with the hard tangible linkage of the coaxial cable. That was a fact, from which reasonable men could hope to draw reasonable conclusions and proceed to reasonable recommendations.

But we have not found it quite that simple. For one thing, the conclusions did not at any time sort themselves out in any orderly fashion. One could not, for example, come to terms with the structure that a cable television system might assume until one had some notion of the content—but the content, it turned out, was contingent upon the structure. The provision of services is the outcome of a constant interplay between supplier and user, in which the nature of the service that is offered is affected constantly by the nature of the response; but the provision of services on cable television we possess at this moment provides neither an organized body of suppliers nor an organized body of users. The system is such that it can offer, for the first time, general public access to an instrument of mass communications, but simply to state that public access exists for the first time is to confess that there are no benchmarks from which to depart upon an exploration of the consequences.

And even though the technology of cable television is a small matter, it has a capacity to perturb which we could never freely dismiss from our deliberations. Television itself is a major social fact. Even disregarding its status as a major industrial enterprise, there are tens of millions of people who shape in some degree their lives upon the hard fact of television: who look to it as their primary source of entertainment, their primary source of news and opinion, their guide if not their determinant when economic and political decisions are to be made. For better or for worse, television as it now exists is a system-in-being of major consequence to those that it serves. When the stakes are that high — and we speak here of the social stakes, not the economic stakes — one does not tinker lightly, and without some sense of apprehension.

The very size and the impressive stability of the existing television system led also to recurring uncertainties about the reality of the exercise in which we were engaged. That cable television had the capacity to perturb existing arrangements we had no doubt. But capacity, of itself, is somewhat an abstract notion. There are technologies by the dozens that lie unused, or only partially used, simply because employing them appears from the near side of the undertaking to be forbidding, or simply because no one wishes to face up to the inconveniences of the period of transition.

Cable television may be one such technology. The current growth rate of 22 percent annually may quite possibly be a gross deception. Certainly it cannot be maintained unless cable makes great strides in penetrating the medium-size and larger cities. To be sure, regulatory decisions have been such as to inhibit up to now that penetration, but there is no assurance that even when the regulations change the penetration will take place. The growth rate may dwindle, and an upper limit of perhaps 20 percent penetration manifest itself; we do not truly believe that such will be the case, but we cannot dismiss the possibility. At 20 percent, very little of what we have projected can occur.

The Promise of Cable Television

Yet we have persisted, and even upon the whole tended to look upon what we consider to be the bright side. We are anything but certain about the odds. But we have been able, at all times, to recognize even if only dimly the enormous rewards that might be consequent upon the successful development of a sizable, viable cable television system.

If one has any faith at all in the value of communications, the promise of cable television is awesome. The power of the existing system is immense; it dwarfs anything that has preceded it. Never in history have so many people spent so much time linked to an organized system of communications. But where it has dominated communications in power, it has been almost trivial in scope. It has dealt primarily with entertainment at a low level of sophistication and quality, and with news and public affairs at their broadest and their most general. It has been obliged to think of the mass audience almost to the exclusion of any other, and in doing so has robbed what it provides of any of the highly desirable elements of particularity.

Cable television is no threat to the power of the total television system. Whatever radiated television can do, cable television can do quite as well. But those characteristics of radiated television that flow from spectrum scarcity need no longer characterize television as a total system, for the television of abundance can offer television the scope it requires to be a complete communications service. The communications system of unrivaled power becomes then a system of unrivaled scope as well, not doing quite the same things as the printing press, doing many things better and a few important things worse, but wholly commensurate with the press.

In our most perplexed moments—and we have had many—we have not lost this vision. We perceive television as a means of improving the human condition, significantly in some instances, perhaps trivially in others, but improving it all the while. We would not willingly give over the attempt to make a reality of what is now only a potentiality.

When we were brought together to undertake this study, we were asked to complete it in a period of roughly eighteen months, and we have done so. In view of the circumstances, the request appears to us to have been reasonable. Whatever may ultimately become of cable television, much of what comes to pass is being currently determined. The Federal Communications Commission is shaping critical rules during these very months. Congress must soon come to terms with copyright legislation: the pressures of new technologies, of which cable television is only one, makes reliance upon 1909 legislation absurd and unworkable. The Office of Telecommunications Policy within the White House is concerning itself with the whole range of communications. It is a period in which steps are being taken, or perhaps worse, omitted, that will determine the manner in which cable television develops and hence its final form. To have been more leisurely about the progress of our own study might well have vitiated any significance it might hope to have.

Issues We Did Not Confront

But a price has been paid. There are important issues this Report does not cover, or merely alludes to in passing. We encountered those issues, recognized their importance, and reluctantly dismissed from consideration those that appeared to us to be off the main highway, or that we feared could not be fruitfully pursued until the passage of time and the accrual of experience had matured them more fully. We can, however, at least pay them the respect of some mention in these concluding remarks.

There are, to begin with, questions that arise out of the possible effects of the fractionalization of audience. Television audiences today go primarily to network affiliated stations; those mass audiences are reduced by only a few percent by independent stations and Public Television. Mass audiences will almost certainly persist in a system shared by cable television, perhaps on four or five networks rather than only three, but there will be many other channels nibbling at the total and dividing among them some small percentage of the audience.

There is a certain social loss here, which has impressed such observers as Eric Sevareid. National cohesiveness is not without its virtues, and that cohesiveness can be stimulated, when the need exists, by access to all the nation at one time. There are virtues in many voices, Mr. Sevareid agrees, but nonetheless there are moments when one voice should be heard.

Even in a cable system, there are recourses. Certainly the President could demand access to the mass channels on cable television, and request that during his address all other channels go dark. But beyond that, there is the fundamental question Mr. Sevareid raises of the effect of fractionization within a medium of communication that dominates all others. It is a deep question that goes ultimately to very fundamental issues; we have not attempted to deal with it in any depth.

We have omitted also the possibility of using the cable system to conduct instant referenda on matters of local, state or national importance. The technological capacity exists, and our own recommendation that digital response systems be provided brings it close to reality. One could, if one wished, shape an entirely new political system around the ability of the individual citizen to make his views known at every moment. Our instincts lead us to believe it would be an extremely bad political system.

But it is not immediately realizable. So long as cable has only limited penetration—even if the limit is quite high—the only capability is for a government run by those who chance to be cable subscribers. That, surely, is totally unacceptable. The issue, in short, does not soon arise, and we chose to ignore it.

As noted in Chapter Nine, we have not chosen to look deeply into the uses of cable television for formal and informal education, for reasons stated in that section of the Report. We feel strongly, however, that such an examination should be made, perhaps when cable television itself is further developed. We believe the potential of cable television within the educational process is immense, but we do not ourselves know how to get a handle on the problems that are raised and we therefore defer to other bodies and other studies.

As this Report notes in Chapter Four, we have been made aware of the broad-scale, long-term social and demographic implications of a pervasive cable communications system, of which cable television, as we have defined it, would be merely a part. The statement made to the Commission on that subject was remarkably impressive, but it is only fair to say that we did not quite know what to make of it in the light of present information and present trends. At best, it was somewhat more visionary than we were prepared to be. Yet it is undeniable that society is in some respects a seamless whole, that communications is an important part of that whole, and that consequently there must almost surely be some interplay between cable communications and, for example, environmental pollution.

Issues of privacy come to mind as soon as one considers cable connections to and from each household. At the very least, there is the possibility of an electronic equivalent of "junk mail" on a scale that the postal service could not hope to match. There is something disconcerting about the notion that someone, somewhere, might know in detail how one chooses to spend a major portion of one's leisure time. We do not believe, however, that these issues are as important with respect to cable television as with respect to the telephone and to data communications, where lines can be tapped with far more serious effects; we leave consideration of such issues to those who are concerned with those fields.

We have dealt not at all with radio, although the capacity of the cable that carries television to carry radio as well is immediately apparent. Except for some incremental implications for most economical use of the electro-magnetic spectrum, we do not believe the matter to be of great significance. Cable would make it possible to provide, in every home, almost limitless radio channels. It makes possible also new kinds of mutually reinforcing uses of television and radio. They appeared to us, without much investigation, to be peripheral. More imaginative students may prove us mistaken.

We have sought to comprehend and to describe the technology of cable, but we did not believe we were constituted properly, as a group, to look more deeply into technology. It is clear that cable television makes possible major technological improvements in television service. With the need removed to detect signals off the air and to discriminate sharply between those signals, the design and construction of television receivers might be simplified and cheapened. Cable makes possible the assignment of one or more wider-band television channels, in which extremely high resolution would be possible; wherever an extremely brilliant picture is of significance, such channels might be most useful. Requirements of compatibility of color transmission with black-and-white reception, which have been built into the existing system and which reduce the efficiency of color transmission, could be dispensed with on cable channels. Indeed, the whole matter of technical standards for transmission requires attention in view of new capacities. We have contented ourselves here merely with the general recommendation that research and development in cable television be encouraged.

We have not concerned ourselves in any depth with purely commercial services, such as data transmission. Others are engaged in making those studies, for there may be major commercial returns to those who develop the best and most rapid services. We are aware that it is a neighboring furrow to the one we have traveled, and in some respects should be ploughed at more or less the same time and with common purposes; we have not attempted to initiate the liaison.

The list of issues unresolved or unmet is incomplete, but it is sufficient to indicate that although we believe we have carried out our own task, we recognize others that remain to be broached. The list makes clear by indirection what we have set out to accomplish. We have tried to establish the prerequisites for a television of abundance and to suggest some of the services that such a television system might provide, including those beyond an enriched provision of entertainment and news. An impressive new instrument of communication has been made available to society by the advance of technology. It remains for society to employ that instrument wisely and well.

SUMMARY OF MAJOR CONCLUSIONS AND RECOMMENDATIONS

The major conclusions and recommendations of the Sloan Commission on Cable Communications are presented here. The chapter of the Report is noted in which a full discussion of the issues involved will be found.

General Conclusions

I. The Commission believes it to be in the public interest to encourage the growth of cable television.

II. The Commission believes that by the end of the decade a cable television system will be in existence which covers 40 to 60 percent of all American television homes; which provides in a majority of instances a capacity of twenty channels and in many instances a capacity of forty channels or more; which possesses a limited capacity for return signals from the home receiver back to the point of transmis-

sion; and which will be extensively interconnected, most
probably by satellite (Chapter Four).

Franchising Considerations

I. The Commission recommends that during the period of the
present decade a limit of ten years be set upon cable television franchises, with provision for purchase at a fair price
of unrenewed franchises (Chapter Twelve).

II. The Commission recommends that franchises specify allocation of channels among various categories of use, including leased channels and public access channels. Minimum standards for such allocation should be prescribed
nationally (Chapter Twelve).

III. The Commission recommends that franchises contain
provisions which will enable the franchising authority to
require the installation of augmented channel capacity, if
it is found necessary (Chapter Twelve).

IV. The Commission recommends that maximum rates to subscribers be set by franchising authorities (Chapter Twelve).

V. The Commission recommends that provision of service
throughout the franchise area be made a condition of
municipal franchises (Chapter Twelve).

Pay Television

I. The Commission concludes that the field of pay television
should be laid open for the entrepreneur (Chapter Six).

II. The Commission recommends legislation that would reserve specified playoff and championship sporting events
for conventional, sponsor-supported television, and would
free other sporting events for competitive bidding between
pay television and conventional television (Chapter Six).

Cable Access to Programming

  I. The Commission believes that competition between the local broadcasting station and the cable system should be managed on the basis of equal opportunity of access to program and to audience (Chapter Five).

  II. The Commission recommends that exclusive program rights within a geographical area be severely limited in time, with somewhat greater latitude permitted for first-run programs (Chapter Five).

  III. The Commission recommends that cable operators in underserved areas be permitted to fill in the blanks in network service, without payment of copyright fees (Chapter Five).

  IV. The Commission recommends that the current policy of requiring cable operators to carry local radiated signals be maintained, without copyright liability (Chapter Five).

  V. With appropriate copyright legislation, the Commission sees no need for special regulations governing the importation of distant signals (Chapter Five).

Ownership

  I. The Commission recommends that network ownership of cable franchises be prohibited (Chapter Twelve).

  II. The Commission recommends that a limit be set on the number of cable subscribers served nationally by any single individual or corporate enterprise engaged in ownership of cable franchises (Chapter Twelve).

  III. The Commission recommends that local commercial television stations and newspapers of general circulation be permitted to seek franchises within their own Standard Metropolitan Statistical Areas, provided no single such enterprise possesses franchises capable of reaching more

than 10 percent of the households within the SMSA, and
provided also that such franchises, in the aggregate, be
capable of reaching no more than 40 percent of all households in the area (Chapter Twelve).

IV. The Commission recommends that Public Television stations be permitted to operate cable franchises without
market restrictions (Chapter Twelve).

V. The Commission recommends that restrictions on commercial stations and newspapers be waived if no other
bidders for cable franchises appear (Chapter Twelve).

VI. The Commission urges that local franchising authorities
give preference for ownership to community non-profit and
profit-making institutions within neighborhoods which have
special social or ethnic problems and needs (Chapter Thirteen).

VII. The Commission believes separation of ownership and programming to be undesirable during the growth period of
cable television, and consequently does not now favor
common carrier operation (Chapter Twelve).

Federal Regulation

I. The Commission urges that the Congress resolve the issue
of copyright by legislation bringing up to date statutes
governing the property right in intellectual products
(Chapter Thirteen).

II. The Commission recommends legislation acknowledging
the power of the federal government to protect the viewing
public if the growth of cable threatens to deprive any significant sector of the public of broadcast service (Chapter
Thirteen).

III. The Commission recommends the establishment of minimum standards for the allocation of channel uses within
cable systems; minimum technical standards; minimum

standards for channel capacity in new systems; non-discriminatory access to leased and public access channels; and rules on ownership of cable franchises (Chapter Thirteen).

IV. The Commission recommends that Congress limit the powers of the state and municipalities to impose a gross receipts tax or similar levy on cable systems (Chapter Thirteen).

V. The Commission recommends that Congress authorize and provide funds for a program, conducted outside the regulatory agency, that would encourage the wise use of cable (Chapter Thirteen).

State Regulation

I. The Commission recommends that each state establish a special agency empowered to direct and regulate the growth of cable television (Chapter Thirteen).

II. The Commission recommends that such state agencies be empowered by legislation to identify appropriate franchise areas within the state; to identify or bring into being officially constituted franchising authorities for each such area; to establish standards above federal limits, where appropriate, for system capacity, technical characteristics, performance and compliance, allocation of channels, non-discriminatory access, duration of franchises, and the like; to require uniform accounting methods; to require that rate-cards, franchise terms and compliance records be deposited with the agency and to make that material generally available; and to receive and adjudicate appeals with respect to cable operations (Chapter Thirteen).

III. The Commission recommends that state agencies of the sort described above be prepared to delegate major portions of their powers to major cities within the state, in

conformity with ordinary "home-rule" procedures (Chapter
Twelve).

General Considerations

I.   The Commission recommends that the fairness doctrine
     not be applied to the operation of public access channels
     (Chapter Eight).

II. The Commission recommends that the equal time rule
     be considered inapplicable to cable television (Chapter
     Eleven).

III. The Commission recommends that the appropriate legisla-
     tive bodies modify existing laws to remove the liability of
     the system operator, as such, for violations of laws con-
     cerning libel, incitement to riot, obscenity, and the like.
     Liability would remain for those who program the channels,
     including the system operator himself to the degree that he
     provides programs (Chapter Eleven).

PRESENT AND PROBABLE CATV/BROADBAND-COMMUNICATION TECHNOLOGY

by John E. Ward

Summary & Conclusions

This report represents the results of a study performed for the Sloan Commission on Cable Communications, the emphasis of which was on technical and cost factors affecting possible growth of present CATV system concepts over the next several decades to encompass additional "wired nation" communications functions, up to and including point-to-point videophone service. Multi-cable, augmented-channel (converter), and switched CATV systems are reviewed, both from a total channel-count viewpoint, and on the basis of two-way capabilities for video and home digital data services.

The major conclusions relative to the questions that were posed by the Sloan Commission as the background for the study are summarized below. Each topic is discussed in detail elsewhere in the report.

Note that in the following all cost figures for distribution cable plant (including subscriber drops) are on a per-subscriber basis, and on the assumption of 100 percent penetration in medium-density residential areas (aerial plant). The relative cost for various systems should remain fairly constant for lower penetrations and/or high-density urban in-

NOTE: Mr. Ward is Deputy Director of the Electronics Systems Laboratory, Massachusetts Institute of Technology. Only the main part of his paper is reproduced here. The complete and unabridged text, including illustrative appendix material, is available free of charge from the Alfred P. Sloan Foundation, 630 Fifth Avenue, New York City 10020.

stallation, with the possible exception of the two present hub-network, switched systems (Dial-a-Program and DISCADE) which run many cables in parallel (the shorter hub-network cable lengths needed in high-density areas may or may not balance the added cost of installing large multi-cable bundles in such areas). Two-way digital service costs are on the basis of the *additional* cost of the subscriber terminal device over and above the cost of the cable plant.

CAPACITY QUESTIONS

A program capacity of 20–26 channels is now available on a single tree-network cable with converters at a per-subscriber cost of $80. Dual-cable VHF-only systems with 24-channel *nominal* capacity (16–20 actual) have a comparable cost ($70), and can be extended to 40–52 channels by adding converters at a total per subscriber cost of $100. The two present hub-network, switched systems have capacities of 20 and 36 channels, with costs of $113 and $186 respectively.

Single-cable converter-system technology appears to limit at about 35 channels ($100), and this capacity may be reached in the next few years (extension beyond 35-channels per cable is clouded by many technical factors). Use of such improved converters in dual-cable plants would yield 70 channels ($120). Costs of doubling the two switched systems to provide 72–80 channels are not as well defined, but would be about $200 and $280 respectively. The greater costs of switched systems should be balanced against other capabilities (see next heading).

The above cable plant costs for tree-network systems are for downstream only. Adding two-way amplifiers costs about 30% more (*plus* terminal costs). Two-way costs of the switched systems will carry a lower increment, but cannot yet be accurately defined.

SWITCHED VS. NON-SWITCHED SYSTEMS (DISTRIBUTION PLUS SOME TWO-WAY)

The conclusions are that switched systems can be roughly competitive with multi-cable or converter systems only at very high penetration levels but that they do completely avoid the present technical problems of distributing large numbers of channels (this may not be of importance, however, as converter system technology improves). The Dial-a-Program system in particular appears to be less flexible from an installation standpoint unless a very high capital investment is made initially to provide 100 percent hookup capability, and except for very large usage of two-way video (which does not appear probable), has no inherent two-way advantage over non-switched systems. In fact, both switched systems now available appear to be less convenient for the more probable two-way data uses over the next 10–20 years (see next heading below).

The future of switched systems will depend on the marketplace and/or the evident or legislated need for their particular capabilities (in severe local-signal areas, point-to-point video, etc.). Costs would rise rapidly, however, in point-to-point use because the present 24- to 336-subscriber, distributed switching centers are too small for efficient point-to-point networking, and larger hubs would be required, with greatly increased cable costs and perhaps individual line amplifiers.

GENERAL COMMENTS ON TWO-WAY CONFIGURATIONS

Physical configurations for obtaining upstream channels include use of sub-band channels on one cable (4 TV channels maximum), a separate upstream cable (up to 35 channels), or in switched systems, one or more per subscriber cable. Any of these are suitable and roughly comparable for all foreseeable subscriber video-origination requirements (note that the total simultaneous upstream TV originations are always limited by the number of available downstream channels where the signals *go*). General point-to-point video services would require a hub-type switched network and a major cost escalation (see "Picturephone vs. Videophone" below).

Modern digital communication technology permits all foreseeable subscriber *data* requirements to be handled very economically in tree-structured systems by time-division-multiplex techniques, requiring only one "channel pair" (a downstream and an upstream channel), and a simple head-end device. Additional capacity, if needed, can be obtained by multiplexing one or two more data-channel pairs. Use of a hub-type switched system actually seems to complicate the problem of providing data services since more hardware (a device per line) is required to gain sequential access to subscriber lines for polling and/or data transmission. A hierarchy of two-way services is presented below.

HIERARCHICAL ORDERING OF TWO-WAY CABLE CAPABILITIES

The following subscriber terminal costs are *over and above* the cable-plant costs given above and do not include the head-end digital computer (a nominal additional capital cost of $2–$10 per subscriber, depending upon the data service requirements):

1. *Simple monitoring*—channel monitoring; simple yes-no buttons; meter reading and alarm systems: terminal cost $50–$100 per subscriber, plus meter upgrading costs.
2. *More general narrow-band communication and control capabilities* —general-purpose keyboards; on-line channel-access control (restricted distribution; pay TV; video tape library services); audio communication (many to one, or party line); limited information retrieval: terminal cost range $125–$250 per subscriber.
3. *General-purpose data capabilities (many to few)*—access to other computer systems for information retrieval, banking, shopping, etc.; electronic mail: terminal cost range per subscriber $250 up to $1,000 depending on desired terminal display and/or hard-copy capabilities, and total data traffic requirements.
4. *Two-way video*—cost range per subscriber from "zero" for portable local origination (assuming two-way cable), to $500 per permanent home-terminal for subscriber origination capability. Total per-subscriber costs for point-to-point switched video services probably range from $2,000–$4,000 for local connection capability only (within a head end), to $15,000 for a national network. (See "Picturephone vs. Videophone" below.)

COMPATABILITY AND STARDARDIZATION

So long as distribution plus the most probable types of two-way data and limited local-origination are the only required cable services, and each operator provides the proper boxes to interface his system to standard TV sets or special cable receivers, all of the different types of distribution systems can co-exist (even with different channel standards), and can still be networked if desired. This compatability can be achieved at each head-end, just as it is now, by suitable channel frequency translations, etc.

Note, however, that the present combination of a subscriber converter and a standard TV set with tuner represents a duplication of function, and that considerable economies could be achieved by large-scale production of special cable receivers: all-channel models in the case of frequency multiplex systems, and simplified "tunerless" models in the case of switched systems. This of course raises the very pregnant question of whether the system or the subscribers should capitalize such special cable receivers, especially since a subscriber would have no guarantee that his special receiver would work on a different CATV system were he to move. Note also that the economies of very large scale can be obtained only if industry-wide standards are adopted for distribution, upstream, and data channels, and for data communication techniques. On the other hand, the technology is evolving so rapidly at present that such complete standardization within the next year or so would certainly be premature. Herein lies the horns of the dilemma.

PRIVACY CONSIDERATIONS

The privacy issue arises only when non-general-access signals exist on the cable, such as Pay-TV, or communications originated by or directed to particular subscribers. A tree-structured cable network has all its frequency multiplexed channels carried into all homes and obviously offers on opportunity (or a challenge) for clandestine monitoring. A hub-structured network, on the other hand, is comparable to the telephone system in this regard—the average subscriber has no ready means of access to circuits or channels other than his own. A brief examination of this question indicates that while cable monitoring is indeed possible in frequency/time-division networks, it requires considerable technical skill, yields dubious benefits (since private signals are likely to be digital and therefore rather cryptic in nature), and can be made as difficult as desired by special encoding techniques as needed.

TELEPHONE VS. CABLE FOR NARROW-BAND DATA

We have carefully examined present experiments in meter reading via telephone, and although no quoted cost figures are available, it seems clear that subscriber devices costing $100–$150 are necessary at present, with possible eventual reduction to perhaps $50. Since a telephone call is involved in gaining access to a subscriber line, exchange switching delays limit the maximum scanning rate to about 300 subscribers per hour, even though a special exchange interface is provided for automated calling by a utility company computer. Also, a meter cannot be read when the telephone is found to be in use. To avoid interference with normal telephone service, the proposal is to read meters only during off-peak hours in telephone usage; thus the actual reading rate will be about 1,000–3,000 per day. The same function can be *added* to a cable two-way data system at a very low incremental cost ($10–$25), with a reading rate of 4,000 or more per *minute*.

PICTUREPHONE VS. "VIDEOPHONE"

The general need or demand for point-to-point visual communication, either on a local or national basis, has not been demonstrated, although the Bell System obviously feels that it will be a viable service eventually. It also represents a capital cost escalation over the present $500–$600 per subscriber voice telephone plant of about 5:1 for the 1-MHz Picturephone (which can use a substantial part of existing Telco "out-of-plant" cable and local switching plant), and perhaps 20:1 for a "start-from-scratch" nationwide 6-MHz service ($12,000 per subscriber). In the latter case, about half the cost ($6,000 per subscriber) would arise in creating local hub-type, 6-MHz distribution and switching networks comparable in function to the 10,000-subscriber local exchanges on which the telephone system is based. Although this is the future role some have suggested for cable systems, these figures make it appear unreasonable to force any provisions in cable systems for accommodating general point-to-point video service. It is concluded that visual point-to-point services are best left to gradual addition of Picturephone facilities by the Telephone Company as the demand for such services actually arises.

Comparison of High-Channel-Count Systems

The purpose of this chapter is to examine the evolving technology of high-channel-count systems, with an eye both to the question of technical factors affecting channel capacity and to the question of relative economics. The main emphasis is on downstream program distribution capability since this seems to be the overriding factor. All foreseeable two-way services can be handled with relatively few downstream and upstream channels.

The three major types of systems to be considered are: (a) *multicable* systems carrying only the standard VHF channels and using standard TV receivers, (b) augmented-channel systems using additional non-standard frequency channels and requiring either a channel converter or a special all-cable-channel receiver for each subscriber, and (c) *switched* systems in which channel selection is performed remotely from the subscriber's premises and either standard TV receivers or simplified "one-channel" designs can be used. Note that systems (a) and (b) can also be used in combination.

A. BACKGROUND

Community antenna (CATV) systems started out as just that—carrying TV and FM-radio programs on the same frequency bands that are used in over-the-air broadcasting, and permitting subscribers to receive these programs on their standard TV and FM receivers without additional equipment. Since the practical maximum frequency that can be carried on a CATV cable (currently 300 MHz) is well below the UHF television band (470–900 MHz), a traditional single-cable system of this type can offer only the 12 VHF television channels which occupy the two frequency bands from 54 to 88, and 174 to 216 MHz. Actually, 12 channels is an upper limit, which is seldom if ever obtained, particularly as CATV systems have departed from their original role of extending TV coverage into remote, marginal-reception areas and become "urbanized"; i.e., moving into areas covered by strong VHF-TV stations. In such areas, direct pickup of broadcast signals in the cable and/or subscriber TV sets creates an interference with cable signals, more fully described in B below, that can make the channels occupied by strong, local stations unusable for cable transmission and reduce system capacity below 12 *usable* channels—to as little as five or six in some top-market TV areas.

At the same time, the saleability of CATV in areas where subscribers can already receive a substantial number of broadcast stations off-the-air was found to depend upon the ability to offer an expanded channel menu. Thus, simply from a marketing standpoint, the CATV industry has for a number of years been evolving ways to regain the lost channels and/or expand somewhat beyond the traditional 12-channel limit, and a few systems currently offer 15–20 channels. More recently, the concept of the "wired-nation" has emerged in which many new uses for cable channels have been proposed, leading to estimated demands ranging from 40 to as high as 60 or more distribution channels and a good deal of controversy about the technical feasibility and economics of such numbers. In addition, various two-way video and digital data communication services are now proposed, leading to further demands on cable technology.

B. TECHNICAL PROBLEMS RELATING TO CHANNEL USAGE

Before getting into the relative characteristics of the three types of systems for providing more than 12 channels, it will be useful to describe one aspect of the technical environment which has influenced their evolution—the group of mutual interference problems that can occur either among cable channels in the process of their transmission to a subscriber and his selection of one of them for viewing, or between cable signals and TV or other signals propagating through the air on the same frequencies. These problems have many complex interrelationships with various system parameters—cable-system distortion factors, receiver and converter characteristics, frequency allocations for non-VHF cable channels, etc.,—and these interrelationships are important to a full understanding of the comparative advantages and disadvantages of the three types of cable distribution systems, and of the current government/industry discussions on cable standards. Since the details are rather voluminous, the problems are simply listed below, together with comments on their relation to the three types of systems under discussion. (A complete description of these problems is given in the full document from which this Appendix is abstracted.)

Several further points should be made concerning the comparison of cable transmission and over-the-air broadcasting in regard to these problems. Two of the six technical problems summarized below (on-channel pickup and cable "leakage") are peculiar to cable systems. The other four problems are almost entirely a function of the characteristics of subscriber receiving equipment (TV sets or converters) and would thus apply equally well to both cable transmission and over-the-air broadcasting, except that the FCC has effectively avoided their effects in VHF-TV broadcasting by a combination of appropriate channel frequency allocations and geographic separation of channel-usage assignments. So long as a cable system carries only the 12 standard VHF channels, it is afforded the same protection against all but one of these latter problems (adjacent-channel interference) so far as receiver characteristics go, although some of the interference effects can be generated within the cable system itself. All the problems apply in augmented-channel systems.

1. *On-channel interference*

Interference at a subscriber's set due to direct pickup of broadcast signals from strong local VHF stations is a problem in all non-switched cable systems; the cures are the shielding of TV sets (impractical for present sets but feasible for new set designs), proper converter design in systems using them (feasible), or a shift to a switched system using only sub-band transmission channels.

2. *Intermodulation and harmonic interference*

   These problems occur primarily in augmented-channel (converter) systems and become more serious the more the number of channels per cable is increased. They are a function of cable frequency allocations, and can arise both in the cable system and the home converters; the cures are better cable amplifiers (feasible) plus proper converter or special cable receiver design (feasible), special choice of cable channel frequencies (difficult) or a shift to VHF-only or to switched systems.

3. *Image interference*

   This is also primarily a problem in augmented-channel systems; the cures are special choice of cable channel frequencies (difficult) and/or proper converter design (feasible), or a shift to VHF-only or switched systems.

4. *Oscillator leakage interference*

   This problem and its cures are roughly the same as (3) for augmented-channel systems. It has not been a problem in single- or multi-cable systems using only the 12 standard VHF channels, even though many TV sets have high leakage.

5. *Adjacent-channel interference*

   This is a problem that affects all non-switched systems and has been particularly troublesome in non-converter systems because many TV sets have poor adjacent-channel rejection characteristics (impractical to fix these existing sets)[1]; the cures are careful control of all cable single levels (already standard practice), better TV set designs for non-converter cable use (feasible but not a short-term solution), proper converter design in augmented-channel systems (feasible), or a shift to a switched system.

6. *CATV "leakage" interference with over-the-air services*

   The problem of signals leaking (radiating) from a cable and interfering with off-the-air reception has been confined in the past to potential interference with non-cable TV viewers because only the VHF-TV frequencies were carried on most cable systems. With the newer augmented-channel systems, there is a potential danger of interference with other non-TV services (such as air-traffic control) in case of a cable break, and the FCC in the future may decide to proscribe certain frequencies from cable usage.

---

[1] In the Malden, Massachusetts cable system an estimated 60 percent of the service calls are for "tuner" problems, and about half of these are adjacent-channel problems, usually leading to TV-set repairs or replacement.

C. MULTI-CABLE SYSTEMS

The first means employed to get around the problem of channel attrition due to direct pickup of broadcast signals was to install two or more parallel cables, each carrying different programs on the same VHF-TV channels. Drops from each cable are brought into each subscriber's home and connected to a switch which permits the subscriber to select one cable at a time for connection to his receiver. This provides a nominal 12 channels per cable, but those channels with direct pickup problems must be omitted from each cable.

1. A Typical Dual-Cable VHF System

As a typical example of this type of system, Table 1 lists the selections available on the dual-cable system operating in Malden, Massachusetts (Malden Cablevision Co.), which include: 15 off-the-air TV stations (nine VHF and five UHF), one local origination channel, two news wires (character displays), and six FM radio stations carried on TV sound channels (for reception through the TV set speaker). Note that since Channels 4, 5, and 7 cannot be carried on-channel in the Metropolitan Boston area because of direct pickup, the Malden system moves these stations to Channels 10, 11, and 12 of cable "A", substituting the FM radio programs on Channels 4, 5, and 7 of both cables. Thus to view Channel 4 (WBZ-Boston), a subscriber must set his cable selector switch to "A" and tune to Channel 10 on his TV receiver. Malden features about eight hours per day of local origination programming (town affairs, school-boy sports, etc.), and this is carried on Channel 13 of both cables.

2. Cost Data

Installation costs for a multi-cable system are of course greater than for a single-cable system, how much more depending upon whether the cables are installed sequentially (upgrading older systems) or all at one time. Double-tracking an existing cable obviously costs as much or more than the original cable since labor and material costs tend to keep rising with time, but two cables can be installed at one time at a cost about 50 percent greater than for a single cable. No exhaustive analysis of cable installation costs has been made, but figures for the Malden system described above appear to be typical for down-stream only plant.[2]

---

[2] Information on the Malden system was obtained in a visit by J. E. Ward, J. F. Reintjes, and R. G. Rausch on January 17, 1971.

TABLE 1. MALDEN CABLEVISION CHANNEL SELECTOR GUIDE

TO TUNE IN:	SET DIAL AT:	TO TUNE IN:	SET DIAL AT:	TO TUNE IN:	SET DIAL AT:
2 WGBH Boston	2A	9 WMUR Manchester	9B	38 WSBK Boston	8B
4 WBZ Boston	10A	10 WJAR Providence	10B	44 WGBX Boston	3B
5 WHDH Boston	11A	11 WENH Durham	11B	50 WXPO Lowell	2B
6 WTEV New Bedford	6B	12 WPRI Providence	12B	56 WKBG Boston	6A
7 WNAC Boston	12A	27 WSMW Worcester	9A	13 MALDEN	13A & B

TV News Services   *FM Stations*

UPI News	3A	WCRB-FM	4A	WHDH-FM	5B
Stock Market	8A	WJIB-FM	4B	WBOS-FM	7A
		WPLM-FM	5A	WEEI-FM	7B

MALDEN CABLEVISION CO. Offices and Studios: 112 Pleasant St. / Malden / Mass. 02148 / (617) 324-0620

NOTES: (1.) Channels 4, 5, and 7 have direct-pickup problems and are carried on cable channels 10, 11, and 12 of cable A. (2.) Channel 13 is a local origination (both cables).

189

The Malden system was installed in 1969/70 by Jerrold Electronics Corporation under a turn-key contract for the entire system. Cable installation costs were not directly broken out, but it is believed that the average cost per mile for 86 miles of dual trunks and feeders (mostly aerial, some underground) was $5,400. The system has an average density of 190 dwellings per mile, and the per-dwelling costs for trunk and feeder installation work out at $28.40, to which must be added $32.00 per subscriber drop (the actual cost in material and labor for installing a dual drop, including pro-rata share of the cable tap). Thus the total per subscriber distribution cost, assuming 100-percent penetration, is roughly $60 for VHF-only carriage (24-channels maximum) without upstream capability.

Although the Malden system can carry frequencies up to 240 MHz, it is operating VHF-only and has not yet used this super-band capability. The addition of converters as described in Paragraph 3 below would permit a substantial increase in downstream channel capacity without any changes in the cable plant. However, retrofitting for sub-band upstream transmission on both cables would add about 30 percent to the trunk and feeder costs, bringing the per-subscriber *cable* cost (100-percent penetration) to about $70.

3. Use of Converters in Multi-Channel Systems

As has been discussed, multi-cable systems operating directly into standard TV sets should provide 12 channels per cable, but lose the same direct-pickup channels on both cables. Thus a more typical capacity in upper-market areas is eight or nine channels per cable, down to as little as five or six in major markets such as New York City and San Francisco. This hardly provides the channel capacity desired for "wired nation" services.

Although converter systems will be discussed more fully in the next section, it should be noted here that their current state-of-the-art is 21–25 nominal channels per cable, thus adding them to a dual-cable system immediately jumps usable capacity to the 42–50 channel region (depending on the type of converter, some of these may be lost due to direct-pickup, harmonic problems, etc.). In fact, this combination seems much more promising for obtaining such capacity than foreseeable extensions of single-cable converter technology, for reasons which will be explained in Section D below. If converter systems improve to 30 or more channels per cable in the future, combination dual-cable/converter systems can then provide 60 channels or more.

The other alternative for very large capacity is a switched system (see Section E).

The addition of converters to a dual-cable system adds about $30 (current price range for 20- to 25-channel converter units) to the other per-subscriber cable-plant costs. Using the figure of $70 per subscriber for a modern two-way, dual-cable plant (from Paragraph 2 above), the total distribution cost of a dual-cable/converter system with upstream channel capability is about $100 per subscriber on a 100-percent saturation basis. This is the figure which is used as a basis for comparison with the single-cable and switched systems.

4. *A Modern System Plan*

A new system being planned for Lawrence, Massachusetts, is an interesting example of current trends, both in high-channel-count and in two-way technology. [3] Initially (for the first 2–3 years), the system will probably operate VHF-only with 24-channel nominal capacity (less perhaps Channels 7, 9, and 11 due to direct pickup). When suitable converters become available, it is planned that one cable will be downstream only and will provide 27 channels (2–13 and A–O) between 54 and 252 MHz. The other cable will provide 13 downstream channels (7–O) and will use a generous frequency band (from 10–108 MHz) for upstream transmission of six video channels and up to 3,500 data channels. Note that with this asymetrical arrangement, only one cable need be retroifitted for upstream transmission. The total of the above is 40 downstream channels, but indications are that it may be necessary to omit perhaps four of the mid-band channels to avoid harmonic problems, leaving an estimated 36 usable channels. Although definite cost figures are of course not yet available, the estimated per-subscriber costs for the eventual configuration are very close to $100, as in Paragraph 3 above.

The Lawrence system is being planned with six completely independent trunks radiating from one head-end, with each trunk having feeder branches in the usual tree-network structure. One advantage of this arrangement is that it multiplies the *simultaneous* upstream capacity by a factor of six — to 36 upstream video channels and up to 21,000 data channels in the probable final configuration. It also offers the possibility of reusing certain channels for different local-interest

---

[3] Information provided by Mr. Thomas G. Polis, Director of Engineering, Greater Lawrence Community Antenna.

programs in each of the six trunk systems. Thus this will be a sort of combination hub/tree network.

D. AUGMENTED CHANNEL SYSTEMS

Use of subscriber converters to solve the direct-pickup problem in cable reception dates from late 1965, when shielded VHF-VHF pre-tuners for the 12 standard VHF channels were first developed by the International Telemeter Corporation as a solution to the direct-pickup problem.[4] The function of these units was to perform the channel tuning in a shielded environment (instead of in the set tuner) and convert to a "quiet" channel (usually 12 or 13) to which the TV set is left tuned. Channels 12 and 13 are the best converter-output channels for a variety of reasons described in Mr. Court's paper, providing alternate output channel choices that are adjacent to insure that one or the other of them will be "quiet" in all areas. Later, converters were constructed with the capability for tuning additional non-VHF frequencies, leading to greater channel capacity per cable.

1. Characteristics of Present Augmented-Channel Converters

The cable industry has generally adopted a set of nine "mid-band" cable channels between 120 and 174 MHz, thus a converter designed to handle these mid-band channels in addition to the standard 12 VHF channels has a maximum capacity of 21 channels. Capacity beyond 21 channels is obtained by also using the "super-band" channels which start at 216 MHz (above Channel 13) and run as high as cable bandwidth will allow. Since few presently installed cables will handle frequencies higher than 240–246 MHz, the current state of the art in tuner-converters is to include the first four or five super-band channels (216–246 MHz), yielding 25- or 26-channel capacity (these numbers were also convenient in converter design; see paragraph 2. below).

Not all converter designs on the market are of tuner type such as described above. One "tunerless" type block-converts seven mid-band channels to high-VHF band (Channels 7–13), with a switch to determine whether the TV set tunes from this group or the standard, un-converted Channels 7–13. Another type is similar in operation but block-converts seven *super-band* channels to Channels 7–13. Both of the above yield a 19-channel capacity. Still another unit block-converts nine mid-band channels to some part of the UHF band for tuning by the

[4] Patrick R. J. Court, "Design and Use of CATV Converters," *Information Display,* March/April, 1971.

TV-set UHF tuner. This latter unit yields 21 channels, but does not provide detent-tuning for the mid-band channels (tuning is the same as for UHF broadcast stations). Since the converted channels are all adjacent, which they never are in off-the-air UHF, one might expect some difficulty in tuning, especially since the selectivity of UHF tuners is usually not as good as that of VHF tuners (actual performance of this unit in practice has not been investigated). All of the tunerless converters that convert to the high VHF band are still subject to direct-pickup problems on the VHF channels, since all tuning is done by the TV receiver.

Table 2 lists some available converters, their conversion schemes, and total channel capacities. The tunerless converters represent a handy solution to the problem of obtaining a few more channels, but will probably disappear in the long run in favor of the tuner type (great-

TABLE 2.   CHARACTERISTICS OF AVAILABLE CONVERTERS
TUNERLESS TYPE (Block Converters):

Mfr.	Model	Conversion	Nominal channels
AEL Comm. Corp.[1]	Superband Tunerless	7 Super-Band to 7-13	19[2]
Tomco Comm., Inc.	SB-1	7 Mid-Band to 7-13	19[2]
Vikoa, Inc.	201M	9 Mid-Band to UHF	21

TUNER TYPE (Conversion to Channel 12 or 13):

Mfr.	Model	Channels			Total channels
		VHF	Mid	Super	
Craftsman Electronic Products Corp.	FV25[3]	1-13	9	4	26
Hamlin Int.	MCC-100	2-13	9	4	25
Tomco Comm., Inc.	Cable-Select II	2-13	9	4	25
TV Presentations, Inc.	Gamut 26	2-13	9	5	26

NOTES: (1) Requires 258-MHz capability. (2) Subject to direct-pickup problems on VHF channels. (3) A push-button tuner with two banks of 13 buttons each (labeled 1-13 and A-M) and a selector switch to determine which bank is active (frequency assignment for channel 1 not given).

er user convenience, more complete solution of interference problems), or new cable receiver designs which tune cable channels directly, eliminating need for a converter at all.

2. *Converter Costs*

Present 25-channel tuner-converter costs are quoted in the range from $30 to $50 depending on the type and manufacturer. Actually, converter technology is in a period of transition. Only a few tens of thousands of converters of all types are yet in use, and although many of these have been cleverly devised from available VHF-TV tuner components which are standard, highly tooled production items, and thus reliable and low in cost, [5] they do not represent an optimum solution to the interchannel interference problems in augmented-channel systems, or to further expansion of channel capacity. The industry is now moving toward new all-channel designs which can take full advantage of interference rejection possibilities and the new extensions in cable bandwidth (to 300 MHz). Until large-scale production builds up over the next few years, such new designs will undoubtedly cost more than present 25-channel designs, perhaps in the $50–60 region.

Assuming that per-subscriber installation cost of a single-cable trunk and feeder system, including drops, is about $50 on a 100-percent saturation basis (based on an assumption of two-thirds of the Lawrence dual-cable system costs given in Section C above), the addition of converters to provide 25-channel capability (or more) brings the total per-subscriber cost of an augmented-channel, single-cable system to $80–$100, depending on actual converter cost.

3. *Ultimate Single-Cable Capacity*

There are obvious limits on the maximum number of downstream channels that can be carried on a single cable:

(a) total cable system bandwidth (largely determined by cable amplifers),

(b) channels which may be unusable for one reason or another (interference effects),

(c) channel needs for upstream transmission.

Within the past six months, cable amplifiers giving good performance

---

[5] For example, at least one of the 25-channel tuner-converters is constructed from two modified 13-position VHF tuner mechanisms in cascade; the first of which tunes the 12 VHF channels and the second, selected by the 13th (UHF) position, tunes 13 mid- and super-band channels. (See the article by P. R. J. Court, cited earlier.)

up to 300 MHz and higher have become available. One series of cable-mounted trunk and distribution amplifiers, offered by Anaconda Electronics, incorporates a postage-stamp-sized hybrid amplifier chip made by the Hewlett-Packard Company. This amplifier chip sells for only $50 and provides excellent performance from 40 to 330 MHz ($\pm$1 dB from linear slope over this range, $\pm$0.3 dB from 40–270 MHz), with very low cross modulation and second-order intermodulation products ($-$89 and $-$80 dB, respectively).[6] Thus 300-MHz bandwidth is now a reality.

The question of how many usable channels can be fitted into the available downstream bandwidth between 40 MHz and 300 MHz on a single cable remains open. At face value the maximum would be $(300 - 42)/6 = 43$ channels, but there are a number of factors which appear to militate against obtaining this exact number:

(a) Possible proscription of some frequencies in the 108–136 MHz region by FCC/FAA as a possible hazard to aircraft navigation/control.

(b) The present VHF Channels 5 and 6, if retained on the cable, don't fit into a contiguous-channel allocation scheme.

(c) Converters to handle such a large number of channels, and cope with all the interference effects possible in a bandwidth of almost three octaves, remain to be demonstrated. Some particular channels may present unsolvable problems and have to be left idle.

It is certain that converter technology will progress beyond the present plateau of 25–26 channels, and may reach 30–35 channels in the near future. How close one can come to completely filling the available bandwidth with usable channels remains to be seen.

E. SWITCHED SYSTEMS

1. Introduction

The previous sections have discussed progress and prospects in the extension of traditional CATV technology, which amounts to broadcasting on a cable network rather than through the air. At least two CATV equipment companies have recently decided that the many complex interference problems encountered in the frequency multiplexing of large numbers of TV channels on an imperfect medium (cable) and the selection among them by subscriber-end "de-multiplexers"

---

[6] HP Specifications "Exhibit C" dated March 15, 1971 for H01-35602A and H02-35602A amplifiers.

(converters and/or TV-set tuners) are better solved, at least in the urban multi-VHF-station environment, by adoption of a completely different approach—remote channel selection in "centralized" switching equipment and transmission of only one (or a few) channels per cable at sub-VHF-band frequencies (less than 50 MHz) selected for minimum interference problems. This of course requires that the subscriber "drops" must be extended to a common point where the selection switching is performed, instead of simply being tapped into the nearest point on a multi-channel, frequency-division-multiplex trunk/feeder system that carries the same signal menu past all subscribers.

The radiating structure of the drops in switched systems characterizes them as "hub networks", as opposed to the conventional CATV "tree network". Since the telephone system as a point-to-point communication service is also a hub network, considerable speculation has developed about this similarity and the possibility of enlarging and/or merging functions, up to and including two-way, point-to-point "videophone" services on the same broadband subscriber cables used for CATV distribution. The intent of this section is not to address this question *per se,* but to compare the emerging switched-CATV systems with the conventional non-switched systems as a means of distributing large numbers of channels. Two-way features and possibilities are also noted. Point-to-point switched video is discussed later.

The two switched systems to be discussed are the Dial-a-Program™ system offered by Rediffusion International, Ltd., and the DISCADE™ (*DIS*crete *C*able *A*rea *D*istribution *E*quipment) offered by Ameco, Inc., both of which started trial operations in mid-1970—Dial-a-Program in Dennis Port, Massachusetts, and DISCADE in Daly City, California, now being expanded into a full-scale 20-channel installation by the system operator. (Announcement was made in October, 1971, that DISCADE has also been selected for a major installation in Salt Lake City.) Because of the important ramifications of this new CATV concept, these systems were thoroughly investigated during the course of the study. Important summary information is presented below.

2. *The Rediffusion Dial-a-Vision System*

The Rediffusion approach is to perform channel selection remotely in switching units capable of handling 336 or more subscribers, and transmit only the one selected channel to the subscriber on a private "drop" wire. A novel feature of the system is the use of a very low frequency distribution channel (3.19 to 9.19 MHz) that can be sent over low-cost twisted wires (instead of coaxial cable) for distances of up to 2,000 feet without amplification. The wires are of the same gauge as the twisted pairs used in telephone systems, but are constructed in a special configuration called a Qwist™ (four wires twisted together). Figure 1 shows two forms of this cable: a single Qwist used for the final drop into a subscriber's premises, and a "6-way" feeder Qwist (six Qwists in one sheath) used to extend groups of subscriber drops to the switching exchange. One wire-pair in each Qwist was de-

FIGURE A-1. QWIST™ DISTRIBUTION CABLE (REDIFFUSION INTERNATIONAL, LTD.)

signed for the program channel, the other wire-pair for switching control. (Rediffusion is currently experimenting with switching control on the program pair, leaving the other pair completely free for other uses.)

The present switching exchange design utilizes rotary, mechanical selectors and permits remote selection of any one of 36 channels. This could be extended to 72 or 108 channels (or more) either by paralleling exchanges of the present design, or by redesigning with larger selectors; in either case without any alteration of the distribution cable network or the subscriber equipment. Thus of all the systems examined, the Dial-a-Program system seems to easily win the channel "numbers game".

In hub networks, total cable-footage required rises much more rapidly as a function of area served than in tree networks, thus there is a dual impetus to keep per-foot cable cost low (as Rediffusion has) and to disperse switching centers to serve relatively small areas. With 336-subscriber exchanges and a 2,000-foot maximum "reach" of subscriber drops, exchange density would be in the range from four to ten per square mile, depending on living density and penetration. In analyzing the cost of a Dial-a-Program installation, the decision was made that the best comparison with non-switched systems would be obtained by considering all costs associated with a single exchange installed in a medium-density single-dwelling area (50 by 100 foot lots) on a 100-percent penetration basis. The analysis includes the exchange equipment plus land and housing for it, the distribution cable network, subscriber equipment, connecting inter-exchange trunk, and a 40-percent allowance for multiple subscribers. (Note that those subscribers with more than one TV set which they wish to use independently require a separate drop and exchange selector.) Head-end costs were excluded, since these are roughly the same for all systems.

The resulting distribution cost for a 36-channel Dial-a-Program system is $186 per subscriber. This is considerably higher than the figures for dual-cable and/or converter systems with comparable channel count, and must be balanced against the additional features of the Dial-a-Program system. The channel extension possibilities have already been mentioned. Another is that there is a ready made, private upstream video channel from each subscriber to the exchange, using 9–15 MHz on the same Qwist pair used for the downstream channel (requires additional equipment at both ends). This permits many simultaneous originations from all points in the system, and could form the basis for a point-to-point, two-way videophone service if additional switching facilities were provided.

Although the Rediffusion subscriber-selector unit ($31) converts the cable channel to Channel 12 or 13 for reception on standard TV sets, it could as easily operate with simplified (tunerless) TV sets by converting to the IF frequency. In fact, Rediffusion estimates 30-percent cost savings for special receivers incorporating their dial-selector unit in place of the usual UHF/VHF tuner. Finally, the complete freedom from all multiplexed-channel interference effects should be considered.

3. *The Ameco DISCADE System*

On close inspection, the DISCADE system is found to be a cross-blending of the traditional tree-structured VHF cable system and a switched, sub-band system as described above. The DISCADE solid-state switching units (Area Distribution Centers) are in smaller sizes (8, 16, or 24 subscribers) and are spliced into aerial or underground cables in the same manner as the usual cable line amplifiers. The network itself is tree-structured, with trunks and feeders (called sub-trunks by Ameco) laid out in the usual pattern. The important difference is that these all have ten coaxial cables running in parallel, each carrying two or four channels frequency multiplexed in a band of frequencies from 5–50 MHz. This provides either 20 or 40 channels total. Because of this low range of channel frequencies, smaller coaxial cables can be used than in VHF or augmented-channel systems, thus cost per mile for the network is not much greater than for a modern dual-cable system (Ameco estimates $11,000 per mile for trunk, $7,500 per mile for sub-trunk, including switch housings, and that the typical ratio of sub-trunk to trunk footage is 10:1).

Each subscriber has a $15 selector-converter unit and single coaxial drop cable (up to 2,000 feet) to the nearest Area Distribution Center (ADC). The largest element of cost is that one 10-pole solid-state switching module must be plugged in to the ADC for each connected subscriber, and these cost $60 each. As in the Rediffusion Dial-a-Program system, multiple TV sets in a single dwelling require separate selectors, drops, and switch modules if they are to be used independently.

The cost analysis of the DISCADE system was carried out under exactly the same assumptions as for the Dial-a-Program system, and resulted in a figure of $113 per subscriber, very close to the figures for dual-cable and augmented-channel systems of comparable channel count. For comparison with the Rediffusion system, the present implementation does not include any upstream channels, nor has a definite plan been stated by Ameco. It is clear that upstream capabilities could be implemented, but no attempt has been made to estimate the costs involved. The network and switching arrangement does not seem to lend itself as easily as the Rediffusion system to any future requirements for point-to-point two-way services. Also, increase in channels beyond 40 would require one additional trunk and sub-trunk cable throughout the system per four channels added, plus either design of new ADC's and larger switching modules, or paralleling of the present units in some way.

On the plus side, DISCADE shares with Dial-a-Program almost complete immunity to the interference effects in the usual multiplexed-channel VHF systems, and the ability to use simplified (tunerless) TV receivers, as in the application of DISCADE at Disneyworld, Florida. Also, the DISCADE system has the advantage that all its equipment is cable-mounted, requiring no real estate to install large switching equipment. Further, it seems more flexible in regard to growth from low to high penetration, since trunks and sub-trunks could be installed throughout an area at about the same per-mile "wire-up" costs as present dual-cable VHF systems (if Ameco's figures are correct), with ADC's being installed only where and when needed to meet actual subscriber hook-up requirements. The Rediffusion exchanges, on the other hand, would have to all (or mostly all) be installed at the outset because they are part of the trunk system, and require ground real estate which would need considerable advance planning and preparation for acquisition, particularly in already settled communities.

The two systems are basically alike in concept, however, and their present technology can and probably will metamorphose as needed to be competitive and to meet any actual or expected demands for two-way data and/or video communications services.

Two-Way Considerations and Systems

A. INTRODUCTION

This section discusses the emerging technology of two-way communication on CATV cables and prospects for the future. Except for a few experimental installations and trials during the past year or so, existing cable plants are only equipped with one-way downstream amplifiers in the trunk lines for the normal TV distribution band from 54 to 216+ MHz. By use of suitable frequency splitters at each downstream amplifier location, it has been found possible to add amplifiers to selectively transmit frequencies below this band (i.e., from roughly 5–40 MHz) in the upstream direction.[7] A number of CATV equipment manufacturers are now beginning to offer such devices, either for new systems or the upgrading of older ones, with the exact frequency range provided varying with the manufacturer.

The availability of say 30 MHz of upstream bandwidth permits up to five 6-MHz upstream channels, some of which can be used for remote TV originations from any point in the system and some for a variety of digital data purposes. Whether five upstream channels is sufficient depends upon what services one wants to provide. To go beyond this, some proposals are to install a separate cable with full upstream bandwidth, use a greater share of one or both cables in dual-cable installations, or as in the Rediffusion Dial-a-Program switched system, provide a separate upstream channel from each subscriber to his program exchange. Whatever the needs, it is clear that substantial upstream cable capacity can be provided within the state of the art and within a factor of two (or less) of the cost of downstream-only configurations.

B. UPSTREAM TELEVISION CHANNELS

The first paragraph below discusses the current trends in the industry; i.e., what is now possible or will be shortly. The second paragraph treats the question of expanding CATV systems to include full two-way, private video transmission on a subscriber-to-subscriber basis and concludes that this would be very expensive and would not fit within tree-structured CATV systems.

1. Present Capabilities

The most obvious use for an upstream TV capability is to permit cablecast originations from any point in the system, transmitting the camera signal back to the head-end or the cable-casting studio for taping and/or live retransmission on a regular downstream channel for general viewing. Needs for such service can probably be satisfied by a few upstream channels on an occasional-usage basis.

The next level of service is to provide certain restricted-access, subscriber-origination video transmission services not connected with general CATV program distribution, such as the interconnection of TV-visual service between schools, municipal or police visual nets, etc.

---

[7] See, for example, H. J. Schlafly, "The Real World of Technological Evolution in Broadband Communications," a report prepared for the Sloan Commission on Cable Communications, September, 1970.

•

This would add to the needs for upstream channels, both in the number of simultaneous channels required, and in average usage. Note also that in order to provide such point-to-point or "one-to-few" services passing through the head-end, a controlled-access downstream channel is required for each upstream channel so used. An interesting example of this class of service is a community conference hookup permitting a controlled group of subscribers (the conferees) to view and participate verbally with their chairman (another subscriber who has a camera and means for controlling viewing access). Such a system is being developed by Vicom Industries, Inc., Dexter, Michigan, and as described in Section C below, they plan to use three upstream and three downstream TV channels (plus additional audio and data channels) to provide for three such conference hookups simultaneously.

2. The Cloudy Future

Beyond the above types of video services that can now or very shortly be offered, there has been speculation about personal point-to-point video services, such as remote medical diagnosis, general "video-phone" service, etc. These of course imply permanent installation of cameras and cable modulators at subscriber locations, which would represent a major cost escalation—by at least $500 per subscriber even when such devices are in large-scale production, and perhaps more. More important from the viewpoint of the cable plant, the number of independent two-way channels needed would be far in excess of the foreseeable extensions of present cable technology, except perhaps for the Rediffusion Dial-a-Program system. Note however that in the Rediffusion system as presently implemented, the individual two-way subscriber lines have a "reach" of only 2,000 feet and the largest hub is 336 lines, which is a rather small base for a generalized point-to-point switching network (the balance of inter-exchange lines to subscriber lines would be very poor). Whether or not the subscriber lines were extended (with two-way amplifiers and/or a change to coaxial cable) to permit larger hubs, total switching gear at least comparable in complexity to the usual 10,000-subscriber telephone exchange would be needed within a typical head-end, since the switching requirements would be the same.

Large-scale point-to-point switching apparatus for 6-MHz channels is probably feasible, particularly if the signals can be handled at baseband-video or at very low carrier frequencies such as used by Rediffusion and Ameco. The Bell System already can switch 1-MHz Picture-phone signals on modified No. 5 crossbar and ESS equipment, and solid-state video crossbar switches (116 by 211 lines) have been constructed for NASA. Large-scale broadband switchgear would have to be developed, however, and would certainly be more costly than present telephone switchgear. Also, interconnections between exchanges would require broadband trunk circuits (in the telephone system sense) which would become extremely costly as the size of the interconnected system expands (one 6-MHz channel occupies the same bandwidth as 1,000 voice-grade telephone channels). Minimum investment costs for the most modern long-haul terrestrial telecommunications systems are $1,750 per channel-mile for 6-MHz television channels (TD-2 Microwave Relay).[8] Present proposals for a domestic satellite (with a backup satellite are based on providing eight 6MHz TV channels (or 10,560 voice channels) at an investment cost of $47 million. Annual lease of two-way TV circuit (two channels) is expected to be $1.8 million, or $5,000 per day (from FCC filings by GT and E/Hughes, 1970).

In this connection, one study by Complan Associates, Inc.[9] has estimated the added capital cost of a complete nationwide 1-MHz Picturephone service serving 100 million subscribers to be $3,000 per subscriber, about five times greater than the investment in the existing voice-grade telephone system, and that a 6-MHz service on the same basis would cost about 1.2 trillion dollars ($12,000 per subscriber). In both estimates, "out-of-plant" and local exchange costs (subscriber lines and terminals, and first-level switching) account for one-third to one-half of the total, thus two-way 6-MHz "videophone" service just within a typical CATV head-end (10,000 subscribers) would cost at least $4,000 per subscriber, perhaps 20 times more than the most probable types of CATV configurations over the next few years.

It should be noted that no analysis comparable to that of the Complan study has been made for generalized point-to-point services during the course of this study. Whether the estimates presented above are correct or not, however, it is clear that the cost multiplier for expansion of CATV systems to include generalized point-to-point switched "videophone" service is quite large and that a hub-type network would be required.

[8] R. D. Swensen, "Investment cost of Terrestrial Long-Haul Telecommunications Facilities," *IEEE Transactions on Aerospace and Electronics Systems,* Vol. AES-7, No. 1, January, 1971, pp. 115–121.

[9] President's Task Force on Communications Policy, Staff Paper I, Part 2, Appendix I, Clearinghouse No. PB 184413, June, 1969.

C. DIGITAL CHANNELS FOR CONTROL, MONITORING, AND DATA SERVICES

There is a wide range of services one can imagine for digital communication via the cable plant. Some of the simpler ones include providing the head-end with information concerning the operation of the cable plant via equipment sensors, or the monitoring of the tuner of each subscriber to gather viewing statistics. Several test installations of this sort are now in progress, and new equipments for these purposes are now coming on the market. Moreover, if current computer communications techniques are employed, it is possible to provide (in order of ascending cost) such services as push-button opinion samples or voting, meter reading, data entry and retrieval from local or remote data banks, electronic mail, and so forth. In the remainder of this section an attempt will be made to examine the technical features and costs of successively more complex systems.

Digital communication on a tree-structured CATV cable is based on the concept of a shared party line, with subscriber stations speaking only when spoken to, i.e., upon receipt of an addressed message. Herein lies one of the interesting problems in devising high-data-rate channels — how to avoid a high overhead in lost channel time due to the variation in round-trip transmission time as a function of subscriber distance, at the same time preventing any possibility of response overlaps. This issue is discussed in the first paragraph below, which includes a proposal for a compensating delay scheme. Subsequent paragraphs discuss the characteristics of three new two-way digital systems, in order of increasing sophistication. The final paragraph discusses the privacy issue.

1. An Approach to Upstream Communication[10]

This note indicates some factors which influence upstream digital data rates for a tree-configured cable communication system. It is shown that data rates close to absolute maximums can be achieved for two different types of systems with very modest equipment. In order to determine data rates it is necessary to assume certain parameters for the system under study. These assumptions are not restrictive since rates which depend on them can be readily scaled to determine the performance of different size systems.

The system under study is assumed to have a single one-megabit-per-second digital channel devoted to upstream communication, as well as

[10] An unpublished memorandum prepared by Professor James K. Roberge, March, 1971.

a downstream channel used to poll the subscriber stations. The average downstream data rate required for polling is low, so this channel can also be used for communication with subscriber stations or for other unrelated functions. The system serves 10,000 subscribers, and the one-way cable distance from the head end to the most remote station is 10 miles.

The simplest system is one where all subscribers are polled sequentially, with a fixed time allotted for reply. This approach limits the maximum upstream data rate of any single subscriber to $10^6 \div 10^4 = 100$ bits per second. This rate is an asymptotic limit which can be closely approached only if some method is included to minimize effects of cable delays.

If no additional delays are introduced at subscriber stations, messages provided in response to polling will be delayed by the round-trip cable transit time between the head end and the subscriber station. This delay depends on the velocity of signal propogation in the cable, and has a maximum value of 150 μs for a 10-mile system using cable with average characteristics. Assuming no compensating techniques are used, this maximum delay represents a dead time overhead which must be included as part of every response interval. The total length of a response interval is dependent on the length of the message transmitted by the subscriber stations each time they are addressed. If a 150-bit message is transmitted by each station, the dead-time overhead reduces the effective average bit rate to 50 bits per second. A message length of 1,500 bits would be required to increase the average bit rate to 90 percent of its asymptotic value, or 90 bits per second. (The above assumes a storage capacity in the terminal for the required message length.)

System complexity can be reduced if some form of compensation is used which reduces the local memory required to achieve a given fraction of the asymptotic bit rate limit. One approach is to adjust the actual polling interval as a function of the distance of the station being polled. The required timing information could be stored in the computer which is presumably located at the head end. This method expands the memory requirements at the head end and also complicates any schemes which time share polling with other functions on a single downstream cable.

A second approach would be to introduce in each subscriber station a compensating delay between the time it detects a poll and the time it starts to respond. Stations located near the head end of the cable would be adjusted for a 150 μs delay, while the most remote stations would have no delay. If delays were correctly chosen, responses would always be detected at the head end 150 μs following polling, and the maximum data rate could be achieved with a constant-frequency, overlapped polling technique. While this method can conceptually be used with no local storage, the minimum practical storage is determined by inaccuracies in the compensating delay. If it is assumed that the delay time can be controlled to within 10 μs (certainly realistic for simple one-shot type circuits), the actual bit rate becomes 99% of the asymptotic value with 100 bits of local storage.[11]

Once the dead-time overhead has been minimized, further increases in capacity are possible only by limiting the number of users who may simultaneously use the system to some fraction of the total number of subscribers. Assume a system which limits the number of simultaneous users to 1,000. Such a system might be organized as follows. A basic time period which allows time for 1,000 equal-length responses is selected. (This time is determined by allowable dead-time overhead and delay time uncertainties as described earlier.) If the system is free of all users, all subscribers are sequentially polled to determine if they wish to get on the system. A complete polling sequence requires 10 basic time periods to poll all 10,000 stations. When station A indicates that it wants regular service, it will be assigned the first response time of each interval, thus assuring uniformly staggered response intervals and an asymptotic bit rate of 1,000 bits per second to station A. Other stations are polled sequentially during the remaining 999 segments of each basic interval, so that slightly more than 10 intervals are required to poll all out-of-service stations. As more stations join the system, they are assigned definite response intervals, and the average rate of polling out-of-service stations decreases. Thus the frequency at which an out-of-service station is polled indicates the degree of loading on the system. When the system is accommodating 1,000 users, it indicates its busy status by never polling out-of-service stations. Stations which cease responding will be returned to out-of-service status to prevent their blocking the system.

Many alternative strategies exist. For example, active stations could receive either more frequent or longer intervals for data transmission when the system is lightly loaded. The disadvantage is that a more

[11] Some people have proposed a cable configuration which would achieve much this same effect. This would be to send responses *downstream* to the farthest end of the cable where they would be returned to the head-end on a separate upstream cable. Signal splitting at the feeder/trunk nodes would be quite complicated, however.

sophisticated subscriber station is required to take advantage of the variable data rate available to it. With the system approach described above, users are assured a known data rate, with uniformly timed intervals for data transmission, whenever they gain access to the upstream cable. If higher upstream data rates are required at any location, multiple stations can be employed.

2. *Vikoa Incorporated*

The system developed by Vikoa appears to have been designed specifically for monitoring the channel being viewed by the subscriber. The system is very slow in comparison to others, requiring 116 miliseconds to poll a single subscriber-response unit for a 5-bit binary number providing for up to 32 different responses. The basic system incorporates a two-level addressing scheme, with 30 groups of 30 subscribers each, or 900 subscribers per outbound command channel. More subscribers can be serviced by frequency multiplexing additional independent command channels. The polling process takes place as follows: the command channel transmits two five-bit characters, using five-frequency tone modulation, to identify the sub-group and subscriber; the next 5 bit times (83.5 ms) are allocated to the response data from the transponder within the subscriber home. This response is transmitted serially using frequency shift keying (FSK) at 60 bits per second.

This system can also be expanded to include alarm monitoring systems and error detection monitors for the cable plant itself. It would be possible to add several bits to the transponder response sequence and thereby interrogate a small set of push buttons on the subscriber unit. At the six-per-second poll rate, however, 1.9 minutes would be required to read the response from all 900 subscribers. This appears to be the limit of the capabilities of the system, and expansion to higher data rates is not possible within this framework. The expected cost of the subscriber unit is $30.00.

3. *Electronic Industrial Engineering, Inc.*

The EIE system has been designed for limited two-way digital communication with as many as 30,000 subscribers, and differs from the other systems described in this section in that the control unit and digital encoder are located at the cable tap—not in the subscriber's home.[12] A dual drop is used: one for video program distribution and one

---

[12] EIE feels that this setup provides for easier servicing and can maintain return-channel security.

for digital signals. Since the unit is designed for use with either converter systems or dual-cable VHF-only plants, it does not include a converter. Currently configured systems include a four-button opinion polling device and a set monitor, but this system can be expanded to include meter reading, cable telemetry, alarm systems and general-purpose keyboards.

The system is designed to use a downstream control channel in the FM band (88–108 MHz) and a return channel at 2 MHz. Each poll consists of a 15-bit subscriber address and a 15-bit command. The response is 128 bits of data which includes the 15-bit subscriber address as an identification. All data transfers are performed on a time-division-multiplex basis which allows 30,000 subscribers to be polled in 22 seconds. Current costs for the subscriber unit are $1,000 per subscriber, but projected costs range from $150 to $200 per subscriber when mass produced.

4. Vicom Industries, Inc.

The most extensive two-way system concept of those discussed here is that being tested and demonstrated by Vicom.[13] The Vicom system is intended to be a complete interactive digital system and has provisions for channel-usage control, data transmission and on-line interaction with the viewing audience (including both keyboard and audio response); all controlled by a computer located at the head-end. Two channels are required to operate the system; the outbound command and control channel operates at a one-megabit rate and transmits 20-bit words at 40,000 words per second to the subscriber terminals. These words are interpreted as either data for a terminal, commands to a terminal, or the polling of a terminal for a response. In any case, the terminal is expected to respond to every poll with its address and a

[13] Details of the Vicom system were obtained in a visit to their plant in Dexter, Michigan, by J. E. Ward and R. G. Rausch on February 25, 1971. We were most cordially welcomed by Mr. Harold W. Katz, President, and all features to be described were demonstrated to us by means of a five-subscriber hookup within the plant. The computer in use was a DEC PDP-8 (about $10,000 plus a special communications interface), but a system using a slightly larger computer was under construction. They estimated computer capital cost at about $5 to $15 per subscriber, depending on the type of data service rendered, based on 4,000 subscribers. Present cost of the home terminal with capabilities as described here is $265, with a projected cost in large-scale production of about $135.

Since the above was written, Telecable, Inc. (a Norfolk, Va.-based MSO) has announced a test of the Vicom terminal in Overland Park, Kansas, with two-way cable facilities installed by EIE. The tests involve home instruction for disabled children, and home shopping demonstrations in cooperation with Sears Roebuck.

single data character, and return response is at the same one-megabit rate.

The home terminal provides (1) basic keyboard entry via 12 momentary contact keys, (2) a microphone for audio responses, (3) a later model will have storage for up to 16 alphanumeric characters to be displayed on the TV-set screen, and (4) a 25-channel converter that may be enabled/disabled under head-end control. Provision is also made for attaching a variety of peripheral devices to the terminal as demanded by the application. These peripherals consist of:

(1) Video cameras
(2) Full-screen alphanumeric generators
(3) Hard-copy printers
(4) Full keyboard typewriters.

A key feature of the Vicom system is the head-end control of who may view any channel. This is accomplished by automatically disabling the converter whenever its channel selector is moved and sending the number of the newly selected channel to the head-end as a response to the next poll. The converter can only be re-enabled by receipt of an enable command from the head-end computer. This is designed for ready implementation of pay-TV and also for controlled-group conferences, as described in the preceding section. The system permits transfer of access-control over a given channel to any subscriber, for instance, the chairman of a conference connection. This control includes the enabling and disabling of the microphones of other subscribers on an individual basis. Thus a person may indicate his desire to speak by pushing a button, but can do so only after the proper enable command is transmitted to his terminal by whoever has control. This control over audio signals is important because the audio channels are handled on a party-line basis.

In order to control the data flow for the programming associated with this equipment, a pair of channels is allocated for the command and response functions. The downstream channel is presently located in the 108–114 MHz band, and the upstream channel is 6–10 MHz. All of the data transmission is on a time-division-multiplex (TDM) basis, and a word consists of 20 one-microsecond bits. Each terminal is addressed at a rate which depends upon the particular application requested by the subscriber, and is controlled by the head-end computer. The rate varies from the order of one per second to five per second for routine polling, and for full-screen display modes, up to 5,000 words per second can be transmitted to a single subscriber (this can be extended to 30,000 words per second if required).

In general, the Vicom system seems to provide a wide range of subscriber interaction and to have a great potential for data transmission both to and from the subscriber. Currently the system is designed for single-cable use and provides 25 forward video channels, 3 reverse video channels, 3 reverse audio channels, and 1 forward and 1 reverse digital channel. This implies that no more than three party line conversations can take place at a time. Vicom stated however that their system could be used with any cable configuration, including the Rediffusion switched system, and channel mix is flexible. The number of party-line interactions is not limited to three by the terminal design.

For the purpose of illustrating the potentials for such a system as that configured by Vicom, consider the following system: 4,000 subscribers and a cable plant which yields a worst-case distance from head-end to subscribers of 5 miles, or a round-trip cable delay of 75 μs. Command and data could be formatted in 20-bit characters as follows:[14]

	1 Bit	12 Bits	7 Bits
Command (1)	1	Address	Command
Data (2)	0	Modifier	data data
		3-bits	8-bits 8-bits

Format (1) above can be used to transmit commands to the subscriber terminal such as: enable channel, set inbound channel A, enable voice, prepare to transmit/receive data, etc. Format (2) above can be used to transmit data to a previously addressed terminal. Assuming that asynchronous communications conventions would be used the total number of bits transmitted per word would be 23; 1 start bit, 20 data bits, and 2 stop bits. With asynchronous conventions, the time between characters is only restricted to being not less than 2 bit times but may be longer to account for cable delays or longer terminal responses.

For the remaining computations we will assume that the home terminals have a built-in delay mechanism such that cable delays may be ignored (except for a 10 μs accuracy limit) and polls sent continuously.[15] Then the transmission of a single poll command would require 23 bit

[14] Vicom considers the details of its message coding proprietary; this is our estimate of how it might be organized.

[15] As described in paragraph 1 above.

times plus 10 μs, or a total of 33 μs for a 1-megabit transmission speed. The resultant poll rate is 30,000 polls per second which could be used to provide a 7.5 character per second data channel to all 4,000 subscribers. If we assume that the background user need not be polled more often than once per second if his set is on and once every five seconds if his set is off, the peak minimum poll rate would be 4,000 subscribers/second with an excess capability of 26,000 polls per second which could be allocated to various subscribers on a demand basis for data entry and retrieval, yielding an ultimate capacity of 50,000 characters per second.

If on the other hand, we assume that cable delays must be allowed for (i.e., that the computer must always wait for a response before sending a new poll), the time per poll must be that for the most distant subscriber: $23 + 77 = 100$ μs. This yields an effective poll rate of 10,000 polls per second; subtracting 4,000 polls as the minimum rate we get an excess capability of 6,000 polls per second to be allocated on demand. Ordering of the poll sequence on the basis of the distance of each subscriber can improve this figure somewhat, and is what Vicom plans to do, but the desirability of a system of eliminating the problems of cable delays should be obvious.

5. *Privacy Issues on the Cable*[16]

Privacy of Communications is a subject that has recently been receiving more attention. Most techniques for digital communications on a cable involve a greater potential for eavesdropping than exists with the telephone system. This is because most CATV systems involve piping all communications going over the cable into the premises of each subscriber where his terminal picks off only those messages intended for him. This means that any subscriber can listen into any other messages by adapting his terminal or replacing it with another one designed for eavesdropping. In a cable digital system, such as those described above, someone who wants to can listen to all the messages on the system with only one device, particularly if there is only one digital channel that is time division multiplexed (TDM) among all the users.

It is not enough however, to say that a cable digital system provides less security than the telephone system. The telephone system does not provide complete security. Wiretapping is not unduly complicated. The relevant question is whether there are a substantial number of people who would undertake the effort needed to eavesdrop on a cable

---

[16] Prepared by Paul J. Fox, graduate student in electrical engineering at MIT.

system but not a telephone system, for obtaining useful information from listening to a TDM digital channel is non-trivial. One has to decode the addressing or timing procedure and other signal protocols. While not overly difficult it does require a serious commitment of time and effort to understand what one is intercepting. There is, however, one fundamental difference between the telephone system and a cable digital system. Any eavesdropping on the telephone system must involve connection somewhere of a physical device which can be detected and found if one is willing to devote enough time and effort to the project. Any subscriber to a cable system does not need a special physical device or connection to listen to his neighbors' messages—only modification of the legally connected device he already has—and even if a special connection *were* used, it would be much more difficult to detect its presence electrically because of the party-line configuration. For those cases in which security of a cable system is important, the digital nature of the messages makes it easy to add cryptographic devices to the system; in fact, this has already been provided for in the Vicom system.

The above remarks of course pertain to a tree-structured cable network. If a hub-type network is employed with space switching, privacy problems are no different from those in the telephone system.

PROSPECTS AND POLICIES FOR CATV

by John J. McGowan, Roger G. Noll,
and Merton J. Peck

The Future Prospects for Cable Television

Without significant penetration in the 100 largest television markets, which contain 87 percent of all television homes, the cable television industry will remain an interesting curiosity of minor economic and social consequence. An appraisal of probable CATV penetration in these markets is therefore essential for assessing the future of CATV, even though such an appraisal must be rather speculative. First, public policy has severely constrained the growth of CATV in the largest markets by prohibiting distant signal importation. As a result, the probable degree of CATV penetration in these markets must be inferred almost entirely from data on the acceptance of CATV in markets with quite different over-the-air viewing alternatives than typically prevail in the 100 largest markets. Second, the regulatory constraints placed on CATV systems in these markets in the future will influence their pros-

NOTE: Mr. McGowan is Professor of Economics, Yale University, Mr. Noll is Senior Fellow, the Brookings Institution, and Mr. Peck is Chairman, Department of Economics, Yale. This study is one in a series called the "Brookings Studies in the Regulation of Economic Activity." The opinions expressed herein are the views of the authors, and do not represent the views of the officers, trustees or staff of the Brookings Institution. Appendix B represents an abridgement of the paper done for the Sloan Commission by Messrs. McGowan, Noll, and Peck. The complete and unabridged text is available free of charge from the Alfred P. Sloan Foundation, 630 Fifth Avenue, New York City 10020.

pective profitability and, hence, the likelihood that CATV service will be offered in these communities.

To deal with the first of these problems the determinants of CATV penetration were estimated from a statistical analysis of a selected sample of large CATV systems. Existing cable television systems provide a means for testing the intensity of demand for different types and numbers of stations. The typical cable television system operates in a locality with few over-the-air viewing alternatives, either because the number of stations in the area is small, or because the topography of the area prevents good signal reception of all or most channels. Cable systems earn revenue by selling access to a larger number of good quality signals. Presumably, the greater the difference between cable and over-the-air options, the higher the price viewers are willing to pay for cable service. Alternatively, for a given price, the more channels offered on the cable in excess of the off-the-air options, the greater the fraction of households abutting the cable that will subscribe. Accordingly, a least-squares regression analysis of the determinants of CATV penetration in a sample of 31 systems, each of which had at least 10,000 subscribers, was performed. The estimated demand equation was then used to estimate the degree of penetration which three hypothetical CATV systems would achieve in markets with various over-the-air viewing alternatives.

System 1 provides minimal service in the larger markets; that is, the only advantage offered is improved reception of existing local signals. In the smaller markets (with fewer than three local network affiliates), the system offers signal quality improvement plus sufficient signal importation to provide three-network service. The estimates indicate that such a system could expect high penetration in smaller markets, but very low penetration in the large markets.

System 2 differs from System 1 in that it provides three network affiliates, all VHF independents receivable off-the-air, a public broadcasting station, plus four more independent stations. (The four extra independents would be either local UHF or imported stations.) In most markets, cable systems would provide three network affiliates, plus four imported independent stations. In medium-size markets with one or more local UHF independents, System 2 would provide fewer than four imported independent signals. In larger markets with some independent VHF and UHF stations, two or fewer signals would be imported. The estimates indicate that System 2 would obtain five or six times as many subscribers in the larger markets as System 1, and nearly three times as many nationally.

System 3 provides three networks, three independent signals in addition to local independent VHF stations, public broadcasting, plus an additional channel offering programs similar to existing network fare in quality and audience appeal. (System 3 is the same as System 2 except that a fourth national network has replaced an independent station.) System 3 would achieve substantially higher penetration than System 2 in markets with the greatest number of over-the-air VHF viewing alternatives.

Table 1 shows estimated nationwide penetration of these three systems. The first two columns show the percentage and number of homes which would subscribe to each system if given the opportunity.

The third and fourth columns of Table 1 show the estimated fraction and number of TV homes likely to subscribe to the various systems, considering that not all parts of the nation are likely to be offered cable service. On the basis of our profitability calculations, cable is unlikely to be installed in areas with fewer than 350 subscribing homes per square mile, which is roughly a population density below 1,000 per square mile.

Roughly ten per cent of American households with television receivers are in rural areas that have population densities under 1,000 per square mile; few of these areas are likely to be offered cable service by private investors regardless of the quality of over-the-air options. In addition, another two per cent of the viewing population lives in urban areas in which more than half of the population is poor and which, therefore, are unlikely to be wired. Of the population living in the top 100 markets, we assume that roughly 15 per cent live in areas with signal reception problems that make cable development profitable despite the number of local over-the-air signals. The remaining popu-

TABLE 1. ESTIMATED NATIONWIDE PENETRATION OF HYPOTHETICAL CATV SYSTEMS

	Homes subscribing if all homes offered the service		Expected households subscribing*	
	(per cent)	(millions)†	(per cent)	(millions)†
System 1	24	14.5	15	9
System 2	64	38.6	48	29
System 3	72	43.0	61	37

*Assumes homes offered service if 350 homes per square mile are likely to subscribe, the minimum size required for an operation that is sufficiently profitable to induce investment.

† Based on a total number of TV households of 60 million.

lation in markets three through 100, leaving out New York and Los Angeles, will subscribe to cable in sufficient numbers to make cable development profitable only if signal importation—System 2—is permitted. Finally, even the two markets with numerous over-the-air options, which contain 13 million TV homes (20 per cent of the national total), will develop cable if cable offers a fourth network-like option.

Based on these assumptions and observations the fraction of households that will be offered cable service will be 20 per cent for System 1. Of the homes offered service, 35 per cent will be outside the top 100 markets; 50 per cent will be in markets three through 100; and 15 per cent will be in the top two markets. System 2 will cause virtually complete wiring of markets three through 100, but no change in the other markets, which means that about 73 per cent of the nation's homes will be offered service. Finally, System 3 leaves only very low-density rural and urban poverty areas unwired; together, these constitute 12 per cent of the population.

Several aspects of the estimates in Table 1 are worth emphasizing. First, in the absence of distant signal importation or some other advantage to stimulate penetration in the 100 largest markets, no more than ten million television homes can be expected to subscribe to cable television. Second, distant signal importation alone is sufficient to change dramatically the likely level of CATV penetration to almost 50 per cent of all TV homes in the country. Third, the estimates suggest that penetration is highly unlikely ever to exceed two-thirds of TV homes. As a result, there will continue to be a substantial public interest in maintaining over-the-air broadcasting; therefore the possible impact of CATV growth on the viability of over-the-air broadcasting cannot be ignored. In sum, without distant signal importation a national cable system will not be developed, and even with distant signals cable will still not become a ubiquitous public utility like power or telephones.

The Impact of CATV Growth on Over-the-Air Broadcasting

Since CATV systems increase subscribers' viewing alternatives, viewing patterns differ between subscribers and non-subscribers. This affects the audience size and advertising revenues of over-the-air broadcasters. In particular, local VHF stations—especially network affiliates—can normally expect to attract smaller audience shares among cable subscribers than among non-subscribers. The effect of CATV growth on UHF broadcasters is more complicated. On the one hand, the quality of UHF signals is equal to that of VHF when both are transmitted over cable. UHF stations should, other things being equal, attract a larger share of the cable audience than of the over-the-air audience. On the other hand, the importation of distant signals works against this effect by dividing the cable audience among more stations.

We have calculated the regression results of a model to predict audience shares of stations. Table 2 uses these results to estimate the effect of System 2 on the local audience for VHF network affiliates, as well as for both VHF and UHF independents.*

In only two instances does System 2 seriously erode local audiences. First, network affiliates in single- and two-station markets suffer audience declines of 50 and 30 per cent, respectively.

Second, VHF independents in the large markets would also experience large local audience losses. At the same time, VHF independents are the stations most likely to be imported into other markets, which would tend to increase the station's total audience. Since the total number of VHF independents in the entire nation is only 19, seven of which are in New York and Los Angeles, nearly all would probably be extensively imported if the limitations on distant signal importation were substantially relaxed or removed. Of course, a local viewer is worth more to the station than a distant viewer, since local advertisers are not generally willing to pay for the latter. A VHF independent may earn anywhere from 15 to 50 per cent of its advertising revenues from local sponsors, with most averaging about 25 per cent.

From the estimates of the effect of more competition on an independent VHF station's audience, the effect of signal importation on advertising revenues can be estimated. Suppose that a distant viewer is only worth two-thirds as much as a local viewer, and that all cable systems (as in System 2) import four independents. An independent that is one of three VHF independents in a market would have no loss in total revenues *if* the number of subscribers to distant CATV systems carrying its signal were one-half the number of homes subscribing to CATV systems operating in its home market. An independent VHF station, operating in a market in which it is the only over-the-air independent, would maintain the same revenues if it were carried on distant CATV systems having a total subscription equal to the homes subscribing to CATV locally.

The estimates presented above are roughly consistent with the

---

* The procedures used to construct these estimates are described in an unpublished appendix to this paper that can be obtained from the authors.

TABLE 2. EFFECT OF PENETRATION OF SYSTEM #2 ON LOCAL STATION'S LOCAL AUDIENCE

	Over-the-Air Alternatives			% Change in Local Audience by Type of Station		
VHF Affiliates	VHF Independents	UHF Independents		VHF Affiliates	VHF Independents	UHF Independents
3	3	1		*	−16.0	+70.0
3	1	1		−3.8	−24.0	+100.0
3	1	0		−3.8	−24.0	n.a.
3	0	0		−7.6	n.a.	n.a.
3	0	2		−7.6	n.a.	+145.0
3	0	1		−7.6	n.a.	+145.0
2	0	0		−28.5	n.a.	n.a.
1	0	0		−50.0	n.a.	n.a.

NOTES: n.a. = not applicable.
* = less than −1.0 per cent.

218

observed results of signal importation in two California cities, Bakersfield and San Diego. In both cities cable systems import seven Los Angeles stations—three affiliates and four VHF independents.

The available evidence, while tentative at best, indicates that network affiliates might lose as much as 15 per cent of their audience to imported signals, but most likely around ten per cent. Since half of the homes in System 2 subscribe to cable, this means a national loss in network audience—and consequently of advertising revenues—of somewhere between four and seven per cent. Since profits as a percentage of sales for networks and affiliates taken together are normally between 15 and 20 per cent, signal importation should cause network system profits to fall by at least 20 per cent, and by at most 50 per cent. For the stations in wired areas with extensive signal importation, profits will fall by more than this. The average VHF affiliate earns profits much higher than the networks, being equal to 25 to 30 per cent of sales. Between one-fourth and one-half of these profits will be erased if the station operates in a market with no independents, and if 80 per cent of the homes subscribe to signal-importing cables (without network duplication); however, these affiliates will still earn profits as a fraction of capital investment that exceed the national average for all types of businesses.

The situation for UHF stations is more serious. The financial position of UHF stations is generally precarious. Half of the UHF network affiliates and 96 per cent of UHF independents lose money. Cable development is likely to be most extensive in the areas served by UHF affiliates, and many of these stations would be severely damaged financially by a ten per cent decline in audience and advertising revenues. For the UHF independents, the picture is so bleak that even a doubling of audience will not pull many stations into the black. Most of these stations now operate only a few hours a day; costs would be much higher if a full broadcast day were attempted. A substantial increase in revenue would be necessary to make these stations financially secure enough to provide services anywhere approaching the role originally envisioned for them by the FCC when the UHF frequency allocation was made 20 years ago.

The policy response contemplated by the FCC is to limit distant signal importation to protect UHF affiliates in small markets and UHF independents in the medium-size markets where most of them operate. After a few years, when all operating television sets are capable of receiving UHF and when further cable development in some areas improves the quality of the received UHF signal, the FCC hopes that the audience for UHF television will rise enough to make UHF viable.

Unfortunately, the prospects for this policy are not bright. More than half of the television receivers in use can receive UHF; a doubling of the audience for UHF, as all sets become capable of UHF reception, simply is not good enough to pull many UHF stations out of the red, particularly in the areas where they are needed most—the middle-to-small markets areas. With or without cable development, UHF's are going to be in financial difficulty for many years. Thus, a signal importation ban would not be likely to provide a sufficient boost to UHF to bring it to the point of fulfilling the role the FCC has in mind, despite sounding the death-knell of widespread cable development in the nation.

CATV Channel Capacity and Supplemental Services

The availability of CATV channels is a key determinant of the number of supplemental services that can be offered. To illustrate this, consider the standard 12-channel system. Seven channels would be required for the three networks and four independent signals. Another channel would be needed for a public broadcasting channel. More channels would be required where over-the-air VHF independents are available. This would leave only two to four channels for supplemental services.

To offer a wide range of services means incurring extra costs for the construction of 20-channel systems. Twelve and 20 are the two choices with the lowest costs per channel, given present technology, although there is some hope of pushing the latter number up to 24 or 25.

Comanor and Mitchell[1] provide perhaps the most detailed cost comparison of 12- and 20-channel systems; their data are the basis for the following conclusions:

1. As between the two principal means for obtaining at least 20-channel capacity—laying two cables or transmitting additional signals on one cable at frequencies adjacent to existing VHF assignments—the latter is considerably cheaper. In addition, the dual-method can result in intercable interference. The fuller use of frequencies around the VHF band requires higher performance amplifiers that add $300 per mile to the cost of distribution cables. Each home set must be provided with a converter which, including installation, costs between $20 and $25.

[1] William S. Comanor and Bridger M. Mitchell, "The Economic Consequences of the Proposed FCC Regulations on the CATV Industry," National Cable Television Association, 1970.

2. The construction of 20-channel capacity including provision of converters to each subscriber would increase capital costs by about 20 per cent over the costs of 12-channel systems. Capital costs are about one-third of total costs, so that additions to capacity increase total costs for the entire system by about seven per cent, or $100 million. The costs per hour of these channels are low. To take a typical example, a 10,000-subscriber system would incur additional costs of about $5 per hour per channel in providing eight extra channels. This is very low compared to the costs of programming and transmission— live programming costs at minimum $50 an hour plus talent costs.

3. The costs for the 20-channel system consist largely of the converters required for each subscriber. It would be possible to construct a 20-channel system and delay the provision of converters until the 20 channels are in use. As 20-channel systems come into use, sets should be manufactured with this capacity just as sets are now manufactured to receive UHF broadcasts. This is particularly attractive because the cost of 20-channel capacity is only about $10 if it is built into the TV receiver at the time of manufacture.

4. The 20-channel capacity on the cable must be constructed at the outset. Converting an existing 12-channel system to 20 channels is extremely expensive; indeed, Comanor and Mitchell report that "the cost of rebuilding an existing system is often somewhat higher than the cost of constructing a new system, since service to subscribers must be maintained during the construction period."

5. Given the major difference in costs of new construction versus conversion it is probably worthwhile to require all new systems to have at least 20 channels. Some operators may have such short time horizons that they will build 12-channel systems.

6. The more visionary uses of CATV require 40 or more channels. Costs begin to rise sharply when multiple cables must be laid or when one cable must be equipped to carry frequencies not around the VHF band (such as, for example, the UHF band). Barring a technological breakthrough, requiring many more than 20 channels in all but the largest systems yields very uncertain benefits at a very high cost.

Subscription Television

A prime claimant for channel assignments on cable systems could be subscription television (STV), or broadcasting available only to viewers paying a fee to broadcasters.

Until recently, STV had largely been written off by most broadcasters and broadcasting regulators as uneconomic. The principal exception was Zenith-Teco, the manufacturer of devices for transmitting and receiving scrambled signals, which has recently made bids to purchase UHF stations in Chicago and Los Angeles on which they intend to broadcast STV. The majority opinion, contrary to Zenith-Teco optimism, is based on the fact that all four of the STV experiments during the past 15 years—three on cable, one over-the-air—financially were either questionable or failures. One STV operation, the Los Angeles system, created sufficient political commotion to cause a state constitutional amendment, later overturned by the courts, to be passed by a referendum of the voters outlawing STV.

Recently, the promise (or threat) of STV has been revived by the prospect that it could be included as one of many services on a cable television system. As only one of many uses of the cable, some of the costs of the system could be shared with other services, improving the chances that STV could prove to be economically viable. This development has not gone unnoticed. Roughly 20 per cent of the local governments issuing cable franchises in the past few months have explicitly forbidden the franchisee to include STV among his services, and last year more than 30 bills were introduced in Congress that would either ban or substantially limit the development of STV.

The point is that the revived interest in STV is justified. If cable is permitted to develop to the extent described in preceding sections, the chances are good that an economically viable STV system can be constructed.

THE DEMAND FOR STV

The indications are strong that consumers would be willing to pay a substantial amount for more viewing options, even when provided a full complement of free network and independent stations. According to the estimates of the demand for viewing options described in the first section, if cable systems provided three networks, at least four independent stations and a public broadcasting station, they could increase the number of cable subscribers by nine million, generating additional revenues of about $800 million, by providing access to another channel with programming similar to that shown by the existing three networks. Even another strong independent station added to the four assumed for System 2 would result in two million more subscribers, or about $180 million additional revenues. These figures provide rough estimates of the amount cable subscribers would be willing to pay for STV.

The Hartford STV experiment in the mid-1960's provides another source of information. Between 1962 and 1964, roughly four per cent of the homes that could receive the signals of the Hartford STV station subscribed to the pay broadcasts. These subscribers paid, on the average, $2.17 a week (or about $113 annually) for STV. In 1971 prices, this figure would be 30 per cent more, or $147 annually. If cable were permitted to develop to the extent predicted by the economic analysis of the preceding sections, reaching 50 per cent of all TV homes, and if four per cent of the homes with cable service subscribed to an STV service charging the same prices as did the Hartford system (corrected for inflation), then the revenues for the subscription would be about $175 million.

A detailed breakdown of the revenues generated by specific program types indicates that the strongest unfilled consumer demand is in program categories similar to those that dominate current network schedules. Movies of recent vintage, boxing and several entertainment categories were the most popular STV offerings; only boxing is not offered regularly on free TV. A fairly recent movie shown three or four times generated revenues of about $550,000 on the scale of a national cable STV system. An average concert produced revenues from all performances of $175,000 on a national scale; an average ballet or opera, $300,000; and a typical nightclub performance, $225,000.

Considering the characteristics of the Hartford STV operation, it did remarkably well. First, Hartford STV was broadcast by a UHF station with all the attendant problems of any UHF. Since, all other things being equal, a UHF station will attract about half as large an audience as a VHF station, STV on cable could be expected to do twice as well. Second, as the only STV station operating, the Hartford STV broadcasters could not afford to purchase programming especially for broadcast on STV, just as no single television station—particularly one capable of reaching a maximum of 100,000 homes—can afford to produce programs of the same quality as network programming. As a result, the Hartford STV station relied primarily on programs produced for other media: recent movies, films of concerts, plays, operas, nightclub acts, and programs shown on free television elsewhere. Third, the Hartford STV station did not devote itself exclusively to STV, instead showing two or three STV programs daily in prime time. The rest of the time was devoted to free, advertiser-supported television; hence the total revenues of the station were higher than those strictly from STV operation. For all these reasons, Hartford did not really offer network-quality programming. Instead, it was more akin to a strong independent, which makes the revenues it generated consistent with the estimates from cable penetration data.

All of these factors, plus the likelihood that the 30 per cent increase in per capita personal income (corrected for inflation) has undoubtedly increased the amount consumers would be willing to spend on STV, suggest that the $175-million, national-market-equivalent annual revenue of the Hartford system is too low an estimate for the revenues that could be generated by a strong, national STV service.

COSTS OF AN STV SYSTEM

To assess the economic viability of STV on cables, the expected revenues must be compared with the expected costs. The most important cost factor is programming expenses. Networks currently spend about $175 million each on program production, including all programs either produced by the network or purchased from independent production companies. These figures are a good place to start in estimating the program costs for an STV system since the hope is that STV programs will be similar in quality to existing network programs.

Programming costs do have considerable flexibility, particularly in the longer run of several years, due to a particular peculiarity of the market for programs. A popular television program is almost impossible to duplicate with a different set of actors and writers. Consequently, the talent associated with a successful program has a position similar to a monopolist's in that it can capture some of the higher profits of successful programs as added income. Successful television performers then earn substantially more than they could earn in alternative jobs (movies, nightclubs, radio, etc.). We have statistically analyzed program costs for all types of network programming for several years in the 1960's and have concluded that if an additional one per cent of the national audience watches a program for a year, then the talent fees increase the following year by about $1,500 more per week than would otherwise be expected. Therefore, the cost of programs for STV will depend to some degree upon the success of the system.

The minimum programming cost is the lowest price that must be paid to induce individuals to work in television rather than elsewhere. Judging from the costs of series that are in their first year or that are marginally successful, these minimum costs are probably under $150 million annually for a full STV system. Of course, for this amount STV could offer full daytime and news service similar to the programming now offered by networks; a prime-time, entertainment-only STV system would cost still less.

For specific program types, the Hartford experiment indicates that movies, boxing, opera, concerts, variety and nightclub programs would all be good candidates to generate revenues more than sufficient to cover costs on STV. This list emphasizes a dual role for STV, one of which generally has not been recognized. First, two of the three "highbrow" program categories, opera and concerts, did rather well in Hartford. One, drama, did poorly (the average drama program, shown three times, generated about $80,000 in revenues on a national scale, which is less than it costs to produce most half-hour situation comedies on free TV), but perhaps drama should not be written off too quickly—the small scale of the Hartford experiment, with its limited resources for programming, may have been especially telling on the attractiveness of the dramatic programming the system was able to present. Filming and broadcasting a single performance of the Metropolitan Opera or the New York Philharmonic—neglecting payments to the performers—would cost on the order of $50,000 to $100,000 (assuming the best sound and picture quality); both organizations could earn revenues two or three times these figures on a national STV system. A series of symphony concerts, featuring three broadcasts each of ten separate concerts of the leading orchestras, could generate revenues in excess of production costs of one or two million dollars, a substantial offset to the deficits of American symphony orchestras.

The second aspect of STV, generally neglected in the discussion of STV, is the overwhelming support for several "low-brow" entertainment categories. The Hartford station, with its low budget, could not experiment with the staple of free TV, the regular series, but all other categories found in the regular TV fare did very well in Hartford, earning revenues that easily would cover production costs.

Another major cost item for networks is payments to affiliates, totalling about $100 million annually. The counterpart in a cable STV system would be payments to cable owners for devoting a cable channel to STV. Historically, broadcasters have not had to pay to have their signals carried by cable systems. One reason for this is that regulation has largely dictated which signals can and cannot be carried. Another reason is that cable owners benefit more from carrying signals than do broadcasters. While the increased penetration of a cable system due to the third network varies according to the other options on the cable and the over-the-air options, on the average the thrid network accounts for about 25 per cent of the cable subscribers. Each cable subscriber is worth $90 in gross revenues, and close to that amount in profits to the cable owner. (The costs of additional subscribers to a given system are very low.) The network captures roughly 25 per cent of the viewing time of all cable subscribers. The network and its affiliate are paid about three cents per hour for every home viewing their signal, so the network system gains about $7 per year for each home on the cable system. The gain of the cable owner from providing the network is perhaps $80 from one-fourth of his subscribers (or $20 per subscriber), which is triple the gain of the network system.

Cable STV would be quite a different matter. As developed herein, an STV channel charging Hartford prices would not lead to much of an increase in penetration (the increase in consumer welfare from one more network channel would be nearly offset by the higher price). The cable system would therefore have to be induced to broadcast STV, particularly since the latter would bear some collection costs. Most likely an arrangement much like that between networks and affiliates would develop between cable owners and STV—the profitability of STV would be split between them; however, this is not a true "cost" to STV, for any payment slightly above collection costs would induce cable owners to carry STV. Thus we shall consider the counterpart to affiliate payments as a residual, part of the STV "surplus" to be divided among the claimants.

The third major expense of STV is transmission costs. If STV were to use the existing long distance microwave links, as do the networks, the costs would be roughly $25 million—about what the networks now pay. STV probably will not pay this much, for the same capacity in a satellite system can be provided considerably cheaper while at the same time reaching all parts of the nation, including areas not served by microwave. While satellite costs can only be estimated with a wide margin of error, a satellite system consisting of a few earth stations that can send signals to the satellite, several thousand stations that can receive signals from the satellite, and three satellite channels would probably have an annual cost of about $13 million. This cost will decline as satellite technology improves and use increases. In addition, cable systems will have additional costs associated with transmitting the received signal; these will total about $5 million dollars annually.

The final expense is the "collection cost"—the extra costs required to prevent viewers from seeing the supposedly pay-broadcasting without paying the fee. The Hartford STV experiment used the most expensive possible technology—broadcasting scrambled signals, and then installing descrambling devices on each subscriber's television receiver. These devices, even if produced in large numbers, would still be likely to cost $100 each. Amortized over a three-year expected life such devices would increase the annual cost of operating a scrambled-signal STV system over cables by about $40 per home, or about $50 million for a system reaching four per cent of the cable subscribers.

The total cost of operating a minimum STV system using scrambled signals would thus be about $200 million, and perhaps less, depending on program costs. Given the demand estimates outlined previously, the system would be very likely to cover costs. And should twice as high a fraction of cable homes subscribe to STV as did to Hartford, costs would increase by only $50 million while revenues would exceed $350 million. This would produce profits exceeding $100 million annually, which is more than are earned by the three existing networks combined.

The STV collection cost could be reduced substantially by making the revenue-collection process more like periodical subscriptions than theatre admissions. This version would collect a regular monthly fee from a subscriber in return for equipping his TV set to receive a particular channel on the cable system. The cost of this system depends to a significant extent on the receiving capability built into the television receiver; however, the cost per receiver as presently constructed for implementing this type of STV service is about $20, assuming mass production. This lowers the total costs of a cable STV system well below $200 million, making such a system clearly viable even at Hartford penetration rates—and with substantial profits.

The type of cable service implied by this STV arrangement is much like the following. For $7.50 monthly (including the amortized installation fee), a subscriber could receive the three networks, several independents (at least four), and public broadcasting. For an additional fee, he could have access to additional service including one channel programmed much like the existing networks or Hartford STV, but free of advertising. A fee of around $12 monthly would, according to the Hartford results, induce at minimum four per cent and most likely more of the homes receiving the first service also to purchase the second. But a more likely result might be that virtually all of the cable subscribers would pay $1 a month for the extra service, generating revenues of $350 million (far above the costs of the system).

PUBLIC ACCESS CHANNELS: THE NEW YORK CITY EXPERIENCE

by Monroe Price and Charles Morris

The Sloan Commission has recommended that one channel in each cable system be reserved for "public access"—or, as the FCC has put it, "that there be one free, dedicated, non-commercial, public access channel available at all times on a non-discriminatory basis." What are the potential uses of a public access channel? What are the barriers to successful use? What needs to be done to see that the concept of public access is meaningfully implemented? The Commission's Report deals generally with these issues; in this paper, we look more specifically at the experience with public channels in New York City, the one metropolitan area where formal public access channels have been in operation.

Since July 1, 1971 the two cable companies franchised in Manhattan have been required to provide two public access channels to be administered in accord with interim regulations promulgated by New York City's Director of Franchises. The regulations represent an important approach to public access channels. To encourage differing uses of the channels, one public channel is an access channel in the best soapbox sense: time cannot be reserved long in advance or for repeated regular use. It is primarily for users who have a one-time or last-minute message which they wish to cablecast. On the second public access channel, there is an opportunity to reserve the same time period each week or several times a week.

NOTE: Mr. Price is Professor of Law, University of California at Los Angeles, and Mr. Morris is the former Research Director of the Center for Analysis of Public Issues, Princeton, New Jersey.

The need for time reservations goes to the heart of the question of the range of uses that may be possible on the public channels. Unless an organization or an individual can be assured of some regularity of appearance on the public channel, the opportunity to develop a viewing "constituency" (a more accurate characterization than "audience") will be slight. The New York City experience demonstrates that for some important uses of the public channels, requiring not insignificant production expenditures, regularity is vital.

On the other hand, providing too great an opportunity for reservation might mean that the public channel would be consumed by a few major non-profit organizations. The New York City interim regulations preclude this danger by imposing limits on the amount of time any one person or organization can take on the public channels. There is a limit of seven hours per week, two in prime time. And reservations can be made for only a thirteen-week period.

Promoting Use of the Public Channels

The New York City experience also suggests the scope of the task of implementing a public channel concept. The basic notion of public access—that persons and organizations in the community have a right to time on a television channel to do, by and large, what they want—is foreign and contrary to experience. Few people seem to realize that public access is really public, and even fewer feel any sense of ease with a medium as highly technical as television.

In New York, several factors were at work which softened the likelihood of underuse of the channels. Shortly before the channels went into operation, the Fund for the City of New York financed a small effort designed to inform organizations about the potential of public channels, to provide technical assistance, to determine through actual production what the costs of public access would be to users, and to monitor municipal administration of the channels. This group, part of the Center for Analysis of Public Issues, remained in existence for the first five months the public channels were in use. Other organizations also sought to create interest in the new cable potential. At New York University, the Alternate Media Center had already amassed great experience in the use of inexpensive video equipment and in community organization. Open Channels, a group which had advocated public access on a federal level, became engaged in New York in testing ways in which public access channels could be best employed.

These promotional agencies were critical in shaping the early uses of the public channels. Despite advance publicity by the cable companies and the press, a sample of 75 likely organizational users of the public channels indicated that none was aware of the existence of the opportunity, and that virtually none had considered using television as part of the way their organization could communicate with their constituency.

The Users

It is far too soon to describe definitively a profile of the users of public access channels in New York City. The channels have been in use only since July 1, 1971. But a pattern is emerging.

1. Special Interest Organizations as Users

In a sense, the most sophisticated development in uses of the public channels has been by specialized organizations which see the channels as an opportunity to reach a narrow constituency effectively. A good example of this narrow-casting approach is the work of the Deafness Research and Training Center.

The Center, a federally funded rehabilitation center, was enthusiastic about the potential of cable and developed an ambitious initial program concept for the city's hearing-impaired population, including a news service, early morning weather and transportation reports, public affairs discussions and entertainment features. This was scaled down to a twelve-week pilot experiment consisting of twenty-four one-hour features shown on prime time. Roughly one show each week is devoted to a taped panel discussion of a topic of particular relevance to the deaf community, while the other features pretaped or prefilmed entertainment features utilizing materials available from, for instance, the National Theatre for the Deaf.

About a month's pre-production time was required to develop lists of films, arrange for studio time to tape the panel discussions, acquire appropriate releases from owners of the film rights, and grapple with the various technical problems that came up. From the start of pre-production, four persons spent about half time on the project.

The technical problems were relatively severe, but most seemed likely to disappear as the cable system developed. The promised interconnection between Sterling and Teleprompter—the two cable systems in Manhattan—has not been effective as of this writing because of a prolonged telephone strike; hence each tape has to be shown separately on both systems. Because neither cable company had one-inch taping equipment compatible with the one-inch equipment at the production facilities (located at Automation House), the bulky Automation House videotape recorder had to be driven to the cable head end for each showing. Bringing in outside recording equipment creates some minor problems at the head end before the equipment is plugged in, warmed up, and the connection effective.

The incompatibility of the two companies' equipment also prevented the Center from showing much of the "canned" material it had planned to use. About the only materials that can be shown by both companies is 16mm film. Interesting educational materials are available on two-inch tape, but on a different kind of tape from that owned by Sterling. The cost of translating one hour of tape to a different tape format or to film is approximately $1,000 per hour, much more than could be afforded.

In the first studio taping, four hours were consumed to produce one hour of tape. Conventional camera angles could not be used for sign language speakers, since the speaker had to be visible full-face from the waist up; positioning the interpreter, and making seating arrangements that permitted effective signing, all took time to work out.

The first panel discussions were somewhat dull. The panel discussion format is dull to begin with, and the necessity for a full-face focus on each speaker eliminates even minimal speaker interaction. Typical subjects have been mental health and the deaf, housing problems of the deaf, and advantages of the English system of signing. It was the original hope not to be restricted to a panel format, but the technical difficulties of mounting anything more ambitious were forbidding. Because of the anticipated intense interest in *anything* available for the signing deaf community, the initial dullness is not a matter of concern, although it will be as the novelty of the programming wears off. The canned entertainment features are, of course, more interesting.

Dr. Jerome Schein, the Director of the Deafness Research and Training Center, has ordered half-inch equipment for the Center and is securing funds for a full-time video staff person to continue the programming. For all his energy, Dr. Schein was no doubt considerably discouraged by the enormous starting problems caused by tape incompatibilities, but the existence of a promotional force providing active technical assistance served to provide needed support. As the programming continues, the acquisition of full-time staff will be critical, since the demands of producing two hours of programming a week are beginning to cut heavily into the Deafness Center's other activities.

Under the auspices of the Center for Analysis of Public Issues, an organization has been specially established to make use of the public channels to reach elderly people, primarily the elderly poor and those who congregate in centers for the elderly in Manhattan. The programs combine entertainment and information and provide programming about the elderly in the City. The tapes include interviews with old people at centers for the aging, a birthday party at an old age center, a discussion of an old people's rights movement, a discussion of legal problems by Jonathan Weiss (a lawyer for the elderly poor), a panel on nutrition, and an exercise program. There is also a taped program on reaction by elderly people to the previous programs.

The organizational work relating to the production for the elderly was formidable. One person, David Othmer, produced and directed the entire operation, requiring about two full months to produce the first ten presentations. The audience as well as the production required organization. But the response of the centers and settlement houses has been uniformly enthusiastic. They feel that the tapes may be an important tool in opening up the sometimes constricted participation of the aged in community life, and are a good way to create a better sense of group and community consciousness.

Ethnic uses are beginning to be more frequent. The Friends of Haiti, a group of exiles, have scheduled regular programs to discuss conditions in their homeland. The Board of Jewish Education is considering a daily pre-school program which will include Hebrew-language instruction. The Chinese Youth Council is actively planning the use of the public channel to assist in job placement, to offer English language instruction, and to help strengthen the community. Plans include providing cable linkage of the one hundred social clubs and two hundred garment factories in the district. In addition to material originated in the community, the Council plans to present television programming which originates in Hong Kong. The major difficulty is that the cable trunk line does not now extend to Chinatown. The Japan Society, an organization dedicated to strengthening communications between Japan and the United States and to increasing appreciation of Japanese culture, is considering regular presentation of videotapes (already produced) dealing with Japanese dance, flower arrangement, Kabuki and Noh. The Italian-American Civil Rights League has reserved time to be used both to reach its members and the community at large.

2. Arts Users

As one might imagine, there is an interesting potential alliance between the public channels and the cultural resources of New York

City. Experiments in Art and Technology, a pre-existing organization, has used the public access channel to nourish a wider audience for artists and to encourage artists to experiment in the use of tape and film. EAT believes that esthetic issues, like social issues, deserve fuller public discussion and that esthetic prejudices have limited the use of television by narrowing ideas about what format, what pace, and what notion of "product" is tolerable on conventional television. The organization has asked thirteen artists to explore the ways in which they would use the medium outside the constraints of conventional television.

The Film Makers Cooperative illustrates another use of public channels by an artistic organization. FMC consists of about 350 avant-garde film makers and has existed primarily to publish a catalogue of current work with the prospect of securing rentals. Since public channels have been available, FMC has presented one of its members, each week, showing his films and discussing his own work.

Under the auspices of Open Channel, the public access channels have been employed to present portions of a medieval festival at the Cloisters, street theater at Lincoln Center, and stories by Diane Wolkstein, the official storyteller of Central Park. There have been tapes, often prepared by groups working with low cost video techniques, on such subjects as the Indian Cultural Center in New York City, an excerpt from a dance performance at the garden of the Museum of Modern Art, and Don Cherry, musician of the streets.

3. Individual Users

This is the most unpredictable class of users. An English professor from New Jersey has lectured on Shakespeare on Saturday afternoons; a critic has presented readings of recent black poetry and criticism. Mrs. Arnez Green has reserved time to present programs of her own creation addressed primarily to black audiences (including profiles of black personalities, and information on community school boards). An ecology expert plans to use the cable channel as a way of communicating with other ecologists throughout the franchise area. Miss Primavera Bowman plans to show her sculpture interspersed with tapes of the sea. Mr. J. C. Thomas, a person who has appeared on commerical television as a critic and comedian, plans a regular thirty-minute show for himself, to provide the freedom and time he feels are denied him on conventional outlets. Mr. Marvin Tabok has been appearing regularly to provide his "Metropolitan Almanac" listing community events each week.

Costs

Public access, if it is to be different from conventional television, implies extremely low cost to users. In New York City, under rates filed with the Director of Franchises, there is no charge to nonprofit organizations for time alone on either franchised system. But cost of time is only a fraction of the costs of conventional programming. Network productions cost in excess of $50,000 per half hour. Community programming must cost next to nothing or be subsidized.

Costs varied in the first few months of public access in New York City. One of the franchise holders, Teleprompter, provides basic studio assistance (one camera and one cameraman) at no charge to non-profit organizations. Thus, users who wish only to employ the simplest production facility can obtain access to Teleprompter's subscribers at zero cost. On the other hand, the rather ambitious project for the elderly, described earlier, cost approximately $5,000 for ten half-hours of programming. The cost of the programming for the deaf was about half as expensive for the same length, primarily because of reliance on studio facilities.

Control of Content

The New York City regulations provide that users must permit the cable company to prescreen programs it plans to present on public channels to determine whether the cablecasting would subject the company to liability. Under the interim rules, the decision of the company is final, except for judicial review, but the company and the user must file a statement of the facts surrounding each dispute with the Director of Franchising. In addition, the user ordinarily is obligated to file an application for time two weeks in advance listing a general indication of the purpose of the program, a statement as to whether any commercial material is included, and a list of individuals who will appear on the program.

Both companies have expressed considerable concern that they risk lawsuits because of the statements of public access users who are irresponsible or simply unaware of the law. The argument that libel and privacy suits are increasingly difficult to win does not impress the companies; they do not fear losing suits, but are concerned with the costs of defense and the danger that a poor experience rating could lead to the loss of insurance or a hefty increase in rates.

Their response for the first few months of public access has been to require prescreening of all material. The rules call for submission of all materials at least two weeks before screening. In fact, both companies have administered the rule broadly, and have been satisfied with submission 48 hours and sometimes even less before air time. To date, at least, there have been no cases of company deletion of offensive material.

The frequently suggested alternative of bonding for public access users to avoid censorship does not appear to be feasible. Contacts with two of the major surety companies in the communication field indicate a total unwillingness to bond uncensored public access users "at any price." They would consider the possibility, although reluctantly, of writing a bond for an umbrella organization, such as Open Channel, to cover all public access users under its aegis. The bond would be written, however, on the assumption that the umbrella organization would have knowledgeable counsel perform the same censorship function currently performed by the companies. Although such an arrangement may be a way around excessive conservatism on the part of the companies, it could also have the reverse effect: since the umbrella organization would be certain to have its bond cancelled after only one or two lawsuits, it might find itself forced to exercise greater caution than the cable companies seem currently disposed to do.

The problem of obtaining bonding for public access users may also have to be forced by proponents of legislation to relieve the cable companies of liability for public access material. It seems quite likely that such legislation would be coupled with a bonding or financial responsibility requirement for users—on the theory that if a user is impecunious or unbonded, and the cable companies are protected from suit by law, a party who has been libeled or otherwise injured over a public access channel would have no recourse at all.

Sterling Manhattan's response to the liability problem has been to require the signing of a very strongly worded indemnification contract, with at least some unusual clauses, such as requiring indemnification for any expenses the company incurs in connection with the cablecast, whether or not the company required the user to join in defense of the suit. Counsel for New York University refused to sign the contract in connection with the productions for the deaf, on the grounds that it was too broadly and vaguely worded. Sterling has not officially consented to showing the material without the contract, but, at the same time, has not interfered with showing of the tapes. Teleprompter has given no indication that it will require a similar contract.

Prospects

At the outset, applications for time on the public channels were few in number. But after the first several months, it appears that there will be demand sufficient to mean that the channels will be used most of the time. The cable companies are already worrying about the problem of choosing among applicants.

Independent centers for use of the public channels are consolidating and seeking a more secure financial base. Some groups, such as the Deafness Research and Training Center, are providing for special departments to concentrate exclusively on the use of cable channels. Open Channel, a demonstration and promotional agency, is establishing local cable committees in the neighborhoods where cable trunk lines are extensive and subscribership high. There is the beginning of entrepreneurship in the public channel area: there are now several video organizations and consultants who seek to provide their services to groups wishing to use the channels.

It is too early for conclusions, but not for some tentative hypotheses.

Public access channels will not be used, generally, to reach the viewing public at large. If a large, distributed audience is desired, public access channels are ineffective. Informing the potential audience of a program can be as expensive as the production itself. Because public access channels will not reach broad audiences in the near future, they are not adequate substitutes for conventional television; they are not yet adequate as a forum for the presentation of competing views on controversial issues.

Public channels will attract and provide a voice for groups who have not gained access to commercial television. But these disenfranchised groups will find that production barriers and difficulties in developing a viewer constituency are often as great as difficulties in obtaining access. Public channels will be used, often, as a kind of closed-circuit medium for internal communication within already established organizations, neighborhoods, and interest groups. If access is the goal, then the cost barriers for most organizations and persons now without access must be extremely low. This includes expenditures for production as well as costs for transmission.

Demand for use of the public channels must be nourished. There will be need for promotional forces within the community and for technical assistance and talent to assist in the preparation of programs. This is especially true during the period of low cable penetration. The cost for each home that views the channel may be high—higher in fact than on commercial channels. Organizations are loath to experiment on public access channels until there is greater subscribership. But without general use, the shape of the public access channel may be skewed and the impetus for their maintenance diminished.

RESEARCH REQUIREMENTS

by J. Herbert Holloman,
J. Francis Reintjes, John E. Ward,
and Jerome B. Wiesner

During the years of transition, when cable grows largely on the basis of its power to provide a more varied diet of entertainment, a strong research and development program should be mounted to provide a better understanding of the capabilities and limitations of existing system configurations, and to develop new system concepts and new configurations. Such an R & D program is essential if cable television is to grow in a fashion that will permit wholly new types of service.

The nature of the particular cable system necessary to permit public-service programming and other non-entertainment applications requires a level of system analysis and technical R & D not likely to occur within the present industry framework. For example, the industry is currently tending in three different directions for the basic cable network:

—multi-cable systems

—single-cable converter systems (with many non-compatible variants)

—hub-network switched systems (several variants)

Each of these has advantages and disadvantages for large numbers of

NOTE: Mr. Wiesner is President, Massachusetts Institute of Technology; and Messrs. Holloman, Reintjes, and Ward are his colleagues at MIT.

downstream channels, two-way communication, privacy, channel-allocation problems, home terminal design and cost, and cable-installation cost. To determine the optimum system, quantitative data are needed to demonstrate the relative merits of these systems in light of various application requirements. Studies of new system concepts and "wired-city" information networks are required to determine the optimum system. New applications of cable television may well require substantially different home terminals. At the early stages before the market is developed, public support for their development will be necessary.

There is the need also for a continuing study of the economics, technical performance and relevancy of service of installed cable systems. In substance, this effort should aim to answer the question: "Are the cable networks really performing in the public interest?" Measurements of signal quality received by subscribers' stations, the monitoring of channel interferences, recommendations for receiver design to minimize channel interferences, recommendations for cost-effective home-terminal design, particularly terminals that meet requirements for new applications (a low-cost terminal that provides hard copies of received information is an example), and a continuing review of cable-network offerings for their responsiveness to the communications needs of the public are all necessary.

This information is required to permit informed decisions of governmental regulatory agencies. Obviously, questions of standardization are involved. The possibilities of new home terminals, for example, may necessitate large-scale production which in turn implies some degree of standardization of channel frequencies, formats and data rates for digital signals, and connection between cable television systems and other information systems.

It is unreasonable to expect that a broad research and development program along lines we envision will in the near future be supported by the private sector. Neither cable operators nor equipment manufacturers have the incentives at this time to support activities that may ultimately alter the format both of the cable system itself and the character of programs provided by the system. The cable-television industry consists of a large number of professional operators and a few suppliers of equipment. As encouragement is given to the rapid spread in use of cable systems, present technology will be employed with large requirements for capital for installation and marketing. Private investment in novel and risky systems is unlikely. The federal government and private foundations are the only possible source of the funds.

With respect to the federal government, we recommend that research and development having to do with the delivery of new services, including the demonstration and development of the technology, should be the responsibility of both the departments of Housing and Urban Development and Health, Education, and Welfare. On the one hand, HUD should increase its support for the development of new and improved cable systems as they relate to the improvement of cities and urban life. HEW should support the experimentation and demonstration for the delivery of information and services related to health, education, and welfare.

The support for the basic technology and for new service development and evaluation should be the responsibility of both the National Science Foundation and the Commerce Department, which apparently has recently been charged with a broad responsibility for communications.

Regulatory information and evaluation of present options should be funded and supported by both the Office of Telecommunications Policy in the White House and the Office of Telecommunications in the Department of Commerce.

We encourage foundations to support particularly those experiments and demonstrations of a highly risky character which are less likely to receive governmental funding and which relate particularly to broad questions of the social costs and benefits of cable television.

It is difficult to estimate precisely the level of research and development activities that ought to be supported by the public or to relate the possible expenditures to social benefits. If cable television is considered to be like most highly technical, electronic, or communications activities, a reasonable level of research and development might be on the order of 60 million dollars a year, about half of which might be carried on by the firms that supply cable television equipment, leaving a program of 30 million dollars as a reasonable level for public support for the next several years.

ABSTRACTS OF COMMISSIONED PAPERS

The following are brief abstracts,
arranged alphabetically by author, of most of
the background papers (in addition to those appearing
in Appendices A, B, C, and D) that were prepared
for the Commission.

Barrett, Marvin — *The Future of News and Public Affairs
Broadcasting and CATV*

Many of the problems that now characterize the broadcasting of news
and public affairs might be solved with cable. Two conditions, however,
have to be met. First, cable needs to offer complete freedom of access,
allowing anyone to make a program or send a message to whatever
audience he can attract. Second, all such programming, with the exception of that which goes over government or public access channels,
should be paid for by the viewer. Such a system, if it included a two-way
capability, might produce a broad band television news service at
comparable to those of newspapers and magazines.

Chayes, Abram — *The Impact of Satellites on Cable Communications*

Communications satellites will be complementary to, not competitive
with, cable systems. Two recent technical developments are of major
importance for the future of satellite communications: random access
(making it possible for any earth station in a satellite system to communicate with any other earth station), and broadcasting capability.
Thus satellites are likely to be the vehicle for large-scale interconnection
of cable systems in the future, and are also likely to be the vehicle that

243

provides broadcast services for homes that are beyond the reach of either cable or terrestrial broadcast systems. Much farther in the future, satellites could become the preferred vehicle for switched interconnection among cable systems.

Cranberg, Gilbert— *Cable Television and Public Safety*

Experiments in the use of cable technology for public safety are now going forward in a few places, such as Liberal, Kansas, and Weston, West Virginia. In larger cities dedicated channels might be used in the future for the training of police and firemen. The cable might also be used for the surveillance of public streets, subway stations, and other areas, as well as for fire and burglar alarms in homes and shops.

Crichton, Judy— *Toward an Immodest Experiment in Cable Television: Modestly Priced*

The problems of ghetto residents as consumers might be alleviated by low-budget programming on the cable. An experiment might involve an overall "key" channel that, like a local newspaper, would provide a large variety of information and news, including school activities, local sports, births and weddings, community meetings, information about nutrition, housing, and shopping, and that would carry the cable-casting schedules of the other local channels. On the other channels might be programs similar to conventional women's programs but aimed at the specific interests and customs of low-income minority groups; programs on money and credit that deal, for instance, with installment buying or with fraudulent sales techniques; programs on medical matters—fees, how to get an eye examination, and where to find a doctor; and programs for shut-ins that tell invalids, old people, and home-bound mothers how to get food, clothes, and services without going out.

Kahn, Ephraim— *Commercial Uses of Broadband Communications*

The immense data-carrying capacity of cable makes it particularly well suited to commercial purposes. Stock market and banking transactions, credit card authorization, retailing, meter reading, market research, and reservation and information services, are a few possibilities among many. Two-way capability, however, is essential to the entrepreneur seeking to sell goods and services via the cable. If two-way capability is not mandated in new cable systems, or does not otherwise become widespread, business will look to competing technologies, especially those offered by the Bell System.

Kalba, Konrad K. — *Communicable Medicine: Cable Television and Health Sciences*

Cable may offer medicine an important vehicle for the continuing professional education of doctors and other health workers, for the remote delivery of health services, and for public education in health. Because cable television is home-oriented, a "health channel" suggests itself, although very little medical programming exists as yet for home consumption. A good deal of audio-visual programming does exist in the area of continuing education for medical personnel; it could easily be put on the cable and thus made available to more practitioners in a more convenient manner than is possible now. Remote diagnosis and consultation are good possibilities for the future, as is educational programming for the general public aimed at prevention.

Kestenbaum, Lionel — *Common Carrier Access to Cable Communications: Regulations and Economic Issues*

Common carrier status for cable means "a framework by which persons desiring to transmit programs or offer services over cable systems are able to do so . . . on a fair and nondiscriminatory basis, without interference or control by the system operator over the user's programs or services." Common carrier status would ease the problem of ownership in cable while at the same time encouraging diversity of programming. Because of its channel capacity, and its ability to add more channels to meet demand, cable lends itself well to use as a common carrier. However, a variety of difficult regulatory, legal, and economic issues have to be resolved.

Mayer, Martin — *Cable of the Arts*

The mass media have always been primarily agents of entertainment in which the arts have led a marginal existence. If the arts are to be resuscitated in cable, the problems of programming, of obtaining an audience, and of paying the bills (the problems of the arts in broadcast television), must be solved. On the matter of meeting costs, the main options are pay television, advertiser-supported programming, and free programming. Pay TV is a good possibility, given an initial break-through in sportscasting. Advertiser-supported programming will depend on the ability of the cable operator to deliver a sizeable audience, which will require some kind of interconnection of cable systems. Free programming in the arts is likely to be esthetically weak and to generate little audience interest. Other basic problems have also to be dealt with, such as the customary "residuals" arrangements, which discourage multiple showings of a given program; the demands of television unions; and in particular the development of new talent.

Mendelsohn, Harold— *The Neglected Majority: Mass Communications and the Working Person*

Special efforts should be made to insure that one or more channels on the cable be managed by, and programmed for, blue-collar workers, whose needs have been neglected by the media in the past. Such channels might provide news and public interest programs aimed at this audience. They might also carry programming that would aid working people in becoming aware of the "outside world as non-menacing," in becoming aware of political demagogery that seeks to exploit blue-collar prejudices and anxieties, and that helps working people with other services such as job training and employment.

O'Brien, Robert E.— *Cable Communications and Social Services*

Because social services involve the dissemination of information, cable may be an important means of carrying such services to a larger number of people, and at lower cost, than is now possible. The cable could be used, for example, simply to tell clients and potential clients what services are available and how to obtain them. It could also carry educational and instructional programming, possibly with videotapes, on such topics as low-budget cooking or job training. If digital feedback is available, cable could help in arranging a job interview or completing a Medicaid form. Child care centers could use specific programming for their children. New stories in large-type displays could be run for clients with sight defects. To realize these and many other possible uses, however, cable penetration would have to be substantial and welfare workers would have to be educated in cable.

Pemberton, John de J., Jr.—*Foreseeable Problems in a System of Maximum Access*

As cable grows, maximum access to channels will become increasingly important. Cable operators should therefore be prohibited from programming altogether, and common carrier status should be mandated for cable. All households in the franchise area should be wired, and subscribers, as well as other programmers, should be given access to channels. Such a system would increase regulatory problems in some respects, such as the control of defamation, fraud, and obscenity; but decrease regulatory problems in other respects, such as maintenance of the fairness doctrine, or imposition of equal time or right of reply rules. Two additional perils—invasion of privacy, and electronic snooping—would require attention.

Price, Monroe E.—*Content on Cable: the Nascent Experience*

Immediately after the FCC announced that large cable systems would have to originate programming (a position later softened and put into question by the courts), there was a great deal of interest in the kinds of programming that cable operators might be able to create. Early plans for such origination ranged from modest efforts to establish community access boards to ambitious ideas for New Town cable systems. In the future, the problem of sustaining the interest of the audience in local origination will remain paramount, whatever the regulatory situation.

Reid, Alex—*New Directions in Telecommunications Research*

Research on the non-technological aspects of telecommunications is needed, especially on two-way, person-to-person systems. The problems of human adaptation to communications technology are important and should be given consideration by policy makers. Measures of effectiveness—comparing the performance of different media for different tasks—should be developed, as should techniques for predicting, and therefore avoiding, the undesirable consequences of communications technology. To date, research on telecommunications systems has been devoted almost entirely to the technology *per se*. Now the non-engineering disciplines—psychology, sociology, management, and others—should be brought to bear on the problems and potential uses of communications systems, including cable.

Ross, Leonard—*The Copyright Question in CATV*

The heart of the copyright question in cable is long-term territorial "exclusivity"; that is, contracts which provide that the program owner cannot license the same show to any television outlet in the geographical area of the purchaser for a specified length of time, usually two to seven years. Such exclusivity arrangements make the most attractive programs virtually unavailable to all but the largest broadcasters in a market area. If cable is to grow, it must have access to programming. Hence, exclusivity contracts should be sharply restricted. They could be restricted in a number of ways—for example, by imposing a short-time limitation, perhaps three months; by compelling subsequent sale of programs on a per-thousand-audience basis; or by prohibiting exclusivity entirely.

Roud, Richard—*Cable Television and the Arts*

At least five types of programming in the arts are possible on cable: 1) artistic performances transmitted from major theaters and centers; 2) original cable and videotape productions, which might help to develop an art form peculiar to cable; 3) instructional films for students of the arts on artistic and craft techniques—for example, violin playing, ballet steps, or pottery making; 4) educational programming for a general audience, particularly in "non-performing" arts such as painting, sculpture, and architecture; and 5) information programs on community arts events.

Schlafly, Hubert, Jr.—*The Real-World of Technological Evolution in Broadband Communications*

The recent explosive development of technologies related to cable communications will shorten the time cycle that in the past has been characteristic of new industries. These technologies are now being applied to the main components of cable systems: the head-end, the cable itself, amplifiers, passive equipment, power supplies, converters, test equipment, etc. Planning for new cable systems will include methods of expanding the number of channels available, channel reuse, two-way capabilities, local distribution service, and other technical developments. Future cable systems may use satellites and lasers, and may include such developments as advanced subscriber response services.

ACKNOWLEDGEMENTS

1. The following persons served as members of the Sloan Commission on Cable Communications:

Edward S. Mason, Chairman, Dean Emeritus, Graduate School of
    Public Administration, Harvard University

Ivan Allen, Jr., former Mayor of Atlanta
John F. Collins, former Mayor of Boston
Lloyd C. Elam, President, Meharry Medical College
Kermit Gordon, President, The Brookings Institution
William Gorham, President, The Urban Institute
Morton L. Janklow, attorney, New York City (Janklow & Traum)
Carl Kaysen, Director, The Institute for Advanced Study,
    Princeton, New Jersey
Edward H. Levi, President, The University of Chicago
Emanuel R. Piore, Vice President and Chief Scientist, IBM
    Corporation
Henry S. Rowen, President, The Rand Corporation
Frederick Seitz, President, The Rockefeller University
Franklin A. Thomas, President, Bedford-Stuyvesant Restoration
    Corporation
Patricia M. Wald, attorney, Washington, D.C. (Center for Law
    and Social Policy)

251

Jerome B. Wiesner, President, Massachusetts Institute of
    Technology
James Q. Wilson, Professor of Government, Harvard University

2. The following persons served full- or part-time on the Commission's staff:

Paul L. Laskin, Staff Director
Monroe E. Price, Deputy Director
Paul W. MacAvoy, Economic Advisor to the Director
Steven R. Rivkin, Counsel
Irving S. Rosner, Technical Advisor

Michael F. Hanson, Staff Assistant
Konrad K. Kalba, Staff Assistant
Sandy Polakoff, Secretary
Carlota Schoolman, Staff Assistant
Mary Schoonmaker, Staff Economist
Gloria S. Scott, Administrative Assistant
Mavis Shure, Secretary
Ralph Lee Smith, Senior Staff Assistant
Christie Post Soutschek, Secretary
John W. Kiermaier — Served as interim staff director while the
    Commission was being formed. Mr. Kiermaier also organized
    special conferences and served as moderator.

3. Background papers, in addition to those done by the staff, were
prepared by the following persons (titles apply as of the time of the
individual's association with the Commission):

John Adler, Partner, John Adler & Associates
Herbert E. Alexander, Director, Citizens' Research Foundation
Marvin G. Barrett, Director, Alfred I. DuPont–Columbia
    University Survey and Awards Program in Broadcast
    Journalism
Abram Chayes, Professor of Law, Harvard University
Gilbert Cranberg, editorial writer, The Des Moines *Register &
    Tribune*
Judy Crichton, free-lance writer
H. Lee Druckman, President, Century Cable Communications,
    Inc., Tucson, Arizona
Ralph J. Garry, Chairman, Department of Curriculum,
    Ontario Institute for Study in Education

Hyman H. Goldin, Associate Professor of Radio and Television,
    Boston University
Ephraim Kahn, free-lance writer and correspondent
John J. Karl, Partner, John Adler & Associates
Lionel Kestenbaum, attorney, Washington, D.C., (Bergson,
    Borkland, Margolis & Adler)
John J. McGowan, Professor of Economics, Yale University
Brenda Maddox, *The Economist,* London
Martin Mayer, author, music critic for *Esquire Magazine*
Harold Mendelsohn, Professor and Chairman, Department of
    Mass Communications, University of Denver
Charles Morris, former Research Director, Center for Analysis
    of Public Issues, Princeton, N.J.
Roger G. Noll, Senior Fellow, Brookings Institution
Robert E. O'Brien, President, Telosco, Inc.
Peter Passell, Assistant Professor of Economics, Columbia
    University
Merton J. Peck, Professor of Economics and Chairman,
    Department of Economics, Yale University
John de J. Pemberton, Jr., Deputy General Counsel, U.S. Equal
    Employment Opportunity Commission, Washington, D.C.
Ithiel de Sola Pool, Professor of Political Science,
    Massachusetts Institute of Technology
Richard Posner, Professor of Law, University of Chicago
Alex A. L. Reid, Project Director, Joint Unit for Planning
    Research, University College, London
Leonard M. Ross, Ezra Ripley Thayer Teaching Fellow,
    Harvard Law School
Richard Roud, Director, Film Society of Lincoln Center; Arts
    Critic, *The Manchester Guardian*
Hubert J. Schlafly, President, Teleprompter Corporation
John E. Ward, Deputy Director, Electronics Systems Laboratory,
    Massachusetts Institute of Technology

4. The following persons appeared as witnesses before the Commission or served in a consulting capacity; many, of course, did both. A number of persons listed in section 3 above as authors of papers also appeared before the Commission but are not listed below. (Titles apply as of the time of the individual's association with the Commission.)

Arthur Alpert, television producer-writer

Harold J. Barnett, Professor of Economics, Washington
　　University, St. Louis, Missouri
Stephen G. Breyer, Professor of Law, Harvard University
McGeorge Bundy, President, The Ford Foundation
Dean Burch, Chairman, Federal Communications Commission
Robert E. Button, Director of Government Relations,
　　Communications Satellite Corporation
Leonard Chazen, member of the New York Bar Association
James Day, President, National Educational Television
Marion Hess Ein, City Hall Complaint Center–Call for Action
　　Program, City Hall, Washington, D.C.
Everett Erlich, Vice President and General Counsel, ABC
Pauline Feingold, Director, Coalition Action Council, New York
　　Urban Coalition
Herman Finkelstein, General Counsel, American Society of
　　Composers, Authors and Publishers
Franklin M. Fisher, Professor of Law, Harvard University
Roger Fisher, Professor of Law, Harvard University
Thomas Freebairn, film producer/director
Fred W. Friendly, Edward R. Murrow Professor of Journalism,
　　Columbia University; Consultant on Television, Ford
　　Foundation
R. P. Gabriel, Technical Director, Rediffusion International, Ltd.
Leonard Goldenson, President, American Broadcasting Company
Peter Goldmark, President, CBS Laboratories
Hartford N. Gunn, Jr., President, Public Broadcasting Service
Barbara E. Harrison, Consultant, St. Barnabas Hospital, New York
J. Herbert Hollomon, Consultant to the President and Provost,
　　Massachusetts Institute of Technology
William K. Jones, member, New York State Public Service
　　Commission
Irving B. Kahn, Chairman, Teleprompter Corporation
John W. Kiermaier, Special Assistant to the Chancellor, Long
　　Island University
Jeffrey B. Kindley, Assistant Professor of English, Columbia
　　University
Richard C. Kletter, Director, Media Access Center, Portola
　　Institute, Menlo Park, California
Kenneth J. Lenihan, Research Associate, Bureau of Social Science
　　Research, Washington, D.C.
Lance Liebman, Assistant Professor of Law, Harvard University

Laura Lippman, student, University of Pennsylvania Medical
School

Paul Webster MacAvoy, Professor of Economics and Management,
Sloan School of Management, Massachusetts Institute of
Technology

Kenneth G. McKay, Vice President, Engineering, AT&T

John W. Macy, Jr., President, Corporation for Public
Broadcasting

Seymour J. Mandelbaum, Assoicate Professor of Urban History,
University of Pennsylvania

Kenneth L. Marsh, co-founder of Peoples Video Theatre, New
York City

Gen. John Martin (Ret.), Assistant Vice President for Domestic
and Aeronautical Satellite Systems, Communications
Satellite Corporation

Gerald Meyer, attorney, New York City (Phillips, Nizer,
Benjamin, Krim & Ballon)

Gerald Phillips, attorney, New York City (Phillips, Nizer,
Benjamin, Krim & Ballon)

John R. Pierce, Executive Director, Communications Sciences
Division, Bell Telephone Laboratories

Vivian Price, student, Graduate Department of History,
University of Texas

Wilbur L. Pritchard, Assistant Vice President and Director of
COMSAT Laboratories, Communications Satellite
Corporation

J. Francis Reintjes, Director, Electronics Systems Laboratory,
Massachusetts Institute of Technology

Eugene V. Rostow, Sterling Professor of Law and Public Affairs,
Yale University

Sol Schildhause, Chief, Cable Television Bureau, Federal
Communications Commission

Theodora Sklover, Executive Director, Open Channel, New York
City

Stanford Smith, General Manager, American Newspaper
Publishers Association

Louis D. Smullin, Professor, Department of Electrical
Engineering, Massachusetts Institute of Technology

Jeffrey S. Stamps, President, Foundation 70, Wellesley,
Massachusetts

Jeffrey Steingarten, associate producer, WNET, New York

Stuart F. Sucherman, Program Officer, Ford Foundation
Joseph C. Swidler, Chairman, New York State Public Service
    Commission
Donald V. Taverner, former President, National Cable
    Television Association
Robert L. Taylor, Publisher, Philadelphia *Bulletin*
Vincent T. Wasilewski, President, National Association of
    Broadcasters
Frank White, Project Director, Foundation 70, Wellesley,
    Massachusetts
Clay T. Whitehead, Director, White House Office of
    Telecommunications Policy

5. In addition to the persons named in sections 2, 3, and 4 above,
the Sloan Commission benefited from the advice and services of many
other people—too many to list here. A number of one-day seminars
were organized by the staff of the Commission on particular subjects
and were attended by numerous persons expert in those subjects. Still
other individuals contributed to the work of the Commission on special
assignments or in response to special requests. The Commission ex-
tends its very sincere thanks to all these persons.

DATE DUE

GAYLORD			PRINTED IN U.S.A.